Germans in the Antarctic

Cornelia Lüdecke

Germans in the Antarctic

Cornelia Lüdecke
Munich, Bavaria, Germany

Translated by Bernard Oelkers

ISBN 978-3-030-40926-5   ISBN 978-3-030-40924-1   (eBook)
https://doi.org/10.1007/978-3-030-40924-1

Translation from the German language edition: *Deutsche in der Antarktis* by Cornelia Lüdecke, © Christoph Links Verlag GmbH 2015. Published by Christoph Links Verlag. All Rights Reserved.

© Springer Nature Switzerland AG 2021

This work is subject to copyright. All rights are reserved by the Publisher, whether the whole or part of the material is concerned, specifically the rights of translation, reprinting, reuse of illustrations, recitation, broadcasting, reproduction on microfilms or in any other physical way, and transmission or information storage and retrieval, electronic adaptation, computer software, or by similar or dissimilar methodology now known or hereafter developed.

The use of general descriptive names, registered names, trademarks, service marks, etc. in this publication does not imply, even in the absence of a specific statement, that such names are exempt from the relevant protective laws and regulations and therefore free for general use.

The publisher, the authors, and the editors are safe to assume that the advice and information in this book are believed to be true and accurate at the date of publication. Neither the publisher nor the authors or the editors give a warranty, expressed or implied, with respect to the material contained herein or for any errors or omissions that may have been made. The publisher remains neutral with regard to jurisdictional claims in published maps and institutional affiliations.

Cover illustration: "Polarsirkel" at Atka-Bay, Antarctica in February 1980. Picture: Oskar Reinwarth, Ottobrunn

This Springer imprint is published by the registered company Springer Nature Switzerland AG
The registered company address is: Gewerbestrasse 11, 6330 Cham, Switzerland

Grottenberg in Kaiser Wilhem II Land, photographed in 1902. (Leibniz-Institut für Länderkunde, Leipzig, Drygalski estate)

Preface

Because of progressive climate change, Antarctica is coming more and more into the focus of public reporting: the ozone hole, calving table icebergs, melting glaciers, and the rising sea level dominate the headlines time and again. The growing tourism in Antarctica increasingly brings people into an actually inaccessible world region, but it also entails problems and hazards associated with shore leave and cruise ship disasters. Antarctica is an object of growing scientific research. To acquire more information for better understanding of the current changes and future climate development, research in the South Polar region at overwintering stations and on research vessels was intensified during the Fourth International Polar Year (2007–2008). However, economic interests in the resources of the continent, which have always played a role in its exploration but may no longer be enforced in accordance with the Protocol on Environmental Protection adopted in 1991, finally call a regulated future in Antarctica into question.

The historiography of German South Polar research began at the same time as the race to the South Pole. As early as 1912, a dissertation discussed Germany's part in solving polar problems, and another doctoral thesis focused on the participation of Germans in exploring the South Polar region during the years from 1901 to 1903.[1] But only after the beginning of Antarctic research in the Federal Republic of Germany in the 1980s did two dissertations focus again on polar history, in which the founding phase of German polar research (1865–1875) and the most significant expeditions from 1900 until World War II were put under study.[2]

---

[1] Rüdiger, Hermann: Deutschlands Anteil an der Lösung der polaren Probleme, Dissertation, in: Mitteilungen der Geographischen Gesellschaft München VII (1912) 4, pp. 455–564; Gerdes, Rudolf: Anteil der Deutschen an der Erforschung des Südpolargebietes, besonders der Westantarktis, bis zur internationalen Erforschung in den Jahren 1901–1903, Dissertation, Borna, Leipzig 1917.

[2] Krause, Reinhard A.: Die Gründungsphase deutscher Polarforschung, 1865–1875, Dissertation, Berichte zur Polarforschung 114 (1992); Lüdecke, Cornelia: Die deutsche Polarforschung seit der Jahrhundertwende und der Einfluß Erich von Drygalskis, Dissertation, Berichte für Polarforschung 158 (1995).

The History of Polar Research Working Group [Arbeitskreis Geschichte der Polarforschung] at the German Society of Polar Research [Deutsche Gesellschaft für Polarforschung] was established in 1991 to satisfy not only general interest in adventures in icy regions but also a professional curiosity.[3] Also, in 2004, the Scientific Committee on Antarctic Research founded the temporary History of the Institutionalization of Antarctic Research Action Group, which was transformed into an Expert Group in 2011.[4]

Against this background, the desire arose to write a popular history of German Antarctic research up until the present day, looking behind the facts known from travel reports and furnished with as many unpublished pictures as possible. The chapters dealing with the first three German Antarctic expeditions are based on my doctoral thesis at the Ludwig Maximilians University and my article on the "Schwabenland" expedition.[5] During the 35th Open Science Conference of the Scientific Committee on Antarctic Research and the International Arctic Science Committee, which took place in Davos (Switzerland) in June 2018, Ursula Rack (an Austrian polar historian working in New Zealand) convinced me to publish an English translation of my book. At the same conference, successful contact was made with Margaret Deignan from Springer Nature Switzerland AG to start this task. For the English translation, provided in a pleasant cooperation with Bernard Oelkers, I have revised, updated, and marginally extended the original text.

We must bear the future in mind to make the right decisions today to ensure that the still mostly untouched continent of Antarctica will not be altered by human impact. To this end, however, looking back into the past is also necessary in order to understand why human beings originally set out on daring expeditions into this inhabitable region and why they later established stations there, which are inhabited all year round.

Initially, whalers and sealers sought to make a fortune on the subantarctic islands since the early nineteenth century and made a lot of money by selling seal pelts, train oil, and the flexible baleens of the whales (whalebone). The scientific exploration of the still completely unknown southern continent started at the turn of the twentieth century, when, during the so-called heroic era, explorers without technical equipment risked their lives to push forward into the continent. But, despite all technological progress, Antarctic exploration is still not a safe adventure today. The contribution made by the German expeditions and research teams in this regard is what this book describes.

Munich, Bavaria, Germany   Cornelia Lüdecke

---

[3] https://www.polarforschung.de/arbeitskreise/ak-geschichte-der-polarforschung/
[4] https://www.scar.org/science/hass/history-group/
[5] Lüdecke: Die deutsche Polarforschung; Lüdecke, Cornelia: In geheimer Mission zur Antarktis. Die dritte Deutsche Antarktisexpedition 1938/39 und der Plan einer territorialen Festsetzung zur Sicherung des Walfangs, in: Deutsches Schiffahrtsarchiv 26 (2003), pp. 75–100.

Acknowledgements

This book was able to be written only with the support of many institutions and archives. Before the reunification of Germany, I had already carried out fundamental research for my dissertation in East and West Germany, where I received much support from the Secret State Archives (in Merseburg and, after reunification, in Berlin), the Central State Archives (later the Federal Archives) in Potsdam, the Federal Archives in Koblenz, and the Archives of the Foreign Office in Bonn (later in Berlin). The Alfred Wegener Institute in Bremerhaven, which is now home to the Archive for German Polar Research, the Federal Institute for Geosciences and Natural Resources in Hanover, the Federal Maritime and Hydrographic Agency in Hamburg, the German Maritime Museum in Bremerhaven, and the Geographical Institute of the Ludwig Maximilians University in Munich also contributed to the book's embellishment with beautiful illustrations.

Above all, however, I benefited from many personal contacts with descendants of expedition members from the time before the Second World War. In particular, I would like to mention Thomas Mörder, Volker Gazert, Gertraude Hartmann, Erich Joester, and Barbara Ronte, whose material I sometimes even had available for decades at home for evaluation. The files, diaries, and other documents handed over by them were a fantastic source, while the photographic collections brought to light many hitherto unpublished pictures. I cannot thank them enough for their trust in my work over many years.

Before the days of Germany's reunification, I found most of the information about Drygalski's South Polar expedition at the Leibniz Institute for Regional Geography in Leipzig, where I was first assisted by the head of the library and archive, Ingrid Hönsch, and later by her successor, Heinz-Peter Brogiato. After the publication of the German edition of this book, I was able to arrange for the bequests of Ernst Herrmann and Alfred Ritscher to be kept by this institute. In Munich, Helmut Hornik, the head of the Filchner Archive at the Bavarian Academy of Sciences, made many unknown photographs accessible to me. I want to thank them very cordially for granting me uncomplicated access to the archives and for their readily given printing permission.

I received photographs of the East German overwintering station and insight into personal experiences from Hartwig Gernandt and Volker Strecke. I got pictures and information about the construction of Georg von Neumayer Station from Klaus-Peter Albrecht and Dietrich Enß. Eberhard Fahrbach documented the dismantling of Filchner Station, and his wife provided me with some photographs after his untimely death.

Georg Kleinschmidt and Hans Oerter told me enthusiastically about their scientific field campaigns and underscored this with their pictures. Also, two overwinterers at Georg von Neumayer Station, Georg Schönhofer and Joachim Schug, as well as one of the station managers Monika Puskeppeleit, shared their pictures and personal experiences with me.

The integration of German polar research into international affairs is documented by photographs from the Americans Paul Bergman and Chuck Kennicutt.

I would like to take this opportunity to cordially thank all of these friends and acquaintances for their support. Without their generosity, I would never have had a look behind the scenes of German Antarctic research after World War II and could never have described and illustrated that time so impressively.

Last, but not least, I thank my editor, Patrick Oelze from the Christoph Links Publishing Company in Berlin, for his very pleasant collaboration in producing the German version of this book.

Finally I received financial support for the English translation from the Goethe Institut, the Neumayer Stiftung, the German Society of Polar Research [Deutsche Gesellschaft für Polarforschung], and the Drygalski Stiftung for which I am very grateful.

Contents

The Race for the Last White Spot on the Map: The First German South Polar Expedition (1901–1903) 1
    The Wilhelminian Policy and the Beginning of Antarctic Research 1
    The Road to Germany's First South Polar Expedition 3
    Preparations for the First German South Polar Expedition 10
    International Cooperation and Other Research Plans 20
    Orders for the South Polar Expedition 27
    The Voyage to the South 32
    Overwintering in the Ice 44
    Back to Cape Town 68
    Homeward Bound 70

Meteorology and Mutiny: The Second German South Polar Expedition (1911–1912) 75
    The Political Framework Conditions for German Polar Research on the Eve of World War I 75
    The Plan for a Second German South Polar Expedition 78
    Training Expedition to Spitsbergen (1910) 83
    Preparations for the Main Expedition 86
    The Departure to Antarctica 93
    In Antarctica 105
    The Return and the Scandal 124
    The Scientific Results and the Aftermath 128
    Excursion: Continuity after World War I—The Foundation of the German Society of Polar Research [Gesellschaft für Polarforschung] 129

The Discovery of Neu-Schwabenland: The Third German South Polar Expedition (1938–1939) 133
    The "Fat Gap", Whaling, and German Possession
    Claims in Antarctica 133
    The Plan for a New German Antarctic Expedition 137
    Preparations for the *Schwabenland* Expedition 145
    The Assignments on the *Schwabenland* Expedition 151
    The Execution of the German Antarctic Expedition (1938–1939) 155
    The Evaluation, New Plans, and Untenable Myths 184
    Excursion: Elements of German Antarctic Expeditions
    Before World War II 189

Separate and United Paths: German Antarctic Research from the End of World War II Until Today 193
    A Private Initiative to Resume Antarctic Research
    in the Federal Republic of Germany 193
    The International Geophysical Year (1957–1958) Without
    German Participation 195
    Antarctic Research in the German Democratic
    Republic Since 1960 196
    Antarctic Research in the Federal Republic of Germany
    Since 1975 201
    The Merging of West and East German Antarctic
    Research After 1990 216

The Future of German Antarctic Research 235
    Further Reading 239

Appendix 241

Picture Credits 263

Abbreviations 267

Chronology of the History of German Antarctic Research 269

About the Author 275

Geographical Index 277

Person Index 281

The Race for the Last White Spot on the Map: The First German South Polar Expedition (1901–1903)

The Wilhelminian Policy and the Beginning of Antarctic Research

In the 1880s, the era of colonial imperialism began for the European powers. Raw materials were to be secured for the motherland, and new sales markets were to be developed for domestic production.[1] For these purposes, they divided the world on the basis of economic considerations. The West Africa Conference (1884–1885) in Berlin, pursuant to its Final Act, was the starting point for partitioning Africa into colonies, and it constituted the German Empire's access to world politics. After an economic crisis lasting from 1882 to 1886, the German Empire took possession of protectorates in Africa, New Guinea, and the South Seas.[2] At that point in time, Germany was developing from an agrarian state into an industrial state and depended on both unfettered access to raw materials and new sales markets.

The industrial boom starting in 1895 and the great expansion of overseas trade with China, Japan, Farther India (now known as Southeast Asia), and trade posts in the Pacific region integrated the German Empire into world economics and world politics.[3] Rear Admiral Alfred von Tirpitz, the state secretary of the Imperial Naval Office [Reichsmarineamt] since 1897, accelerated Germany's fleet-building operations in order to secure overseas interests.[4] Above all, the respect of the greatest sea

---

[1] Messerschmidt, Manfred: Reich und Nation im Bewusstsein der wilhelminischen Gesellschaft, in: Schottelius, Herbert/Deist, Wilhelm (ed.): Marine und Marinepolitik im kaiserlichen Deutschland 1871–1914, Düsseldorf 1981, 2nd edition, pp. 11–33; Westphal, Wilfried: Geschichte der deutschen Kolonien, Munich 1984, pp. 261–265.

[2] Westphal: Deutsche Kolonien, pp. 116–118, 350–351.

[3] See also Witt, Peter-Christian: Reichsfinanzen und Rüstungspolitik 1898–1914, in: Schottelius/ Deist (ed.), Marine und Marinepolitik im kaiserlichen Deutschland, p. 148; Beiträge zur Flottennouvelle 1900, Berlin 1900.

[4] Tirpitz, Alfred von: Erinnerungen, Leipzig 1920, pp. 79–87.

power—England—for Germany was supposed to grow as soon as the German Empire empowered itself to enter alliances against England with a strong fleet of its own.[5]

At the turn of the twentieth century, geographers saw that Germany had also great prospects "of gaining an important position in world trade."[6] Against this trade policy background, legislation was passed to enable gradual further development of the German fleet.[7] Initially, staffing issues were of primary importance.[8] Because of the transition from sail to steam in navigation, there was a lack of qualified seamen. The German Imperial Navy could no longer rely on recruiting trained sailors from the merchant navy or the fishing fleets; it had to train its own crews instead. After the First Fleet Act was passed in 1898, several campaign associations were established within a year, pursuing understanding and approval for the fleet policy among the general population by organizing lectures that argued in favor of the fleet question.[9] The foundation of the Marine Science Institute and Museum [Institut und Museum für Meereskunde] at the University of Berlin stood in the same context.[10]

The fleet budget associated with the fleet acts consisted of a fixed sum of money, which did not leave any margin for unforeseen endeavors, although fleet building under Tirpitz took place during an economic upswing.[11] However, a comprehensive taxation policy did not exist yet; instead, there were only direct tax revenues from the single federal states. In the years 1898 and 1899, fiscal budgets could be balanced only by some unexpectedly high additional revenues resulting from the empire's own tax collections, whereas the next two financial years closed with soaring deficits. In 1902 and 1903, a sound state budget had already ceased to exist. The requirements of the army, the navy, and the East Asian expedition, which founded the naval base of Tsingtau (now known as Qingdao (Shandon Province, China)),[12] had increased considerably; thus, around 1900 and in the following years, military expenditure accounted for about 90% of the imperial budget.[13]

---

[5] See also Böhm, Ekkehard: Überseehandel und Flottenbau. Hanseatische Kaufmannschaft und deutsche Seerüstung, Düsseldorf 1972, p. 89.

[6] Drygalski, Erich von: Deutschlands geographische Lage zur See, in: Beiträge zur Flottennouvelle 1900, Berlin 1900, p. 93.

[7] Compiled from Böhm: Überseehandel, p. 183; Lange, Annemarie: Das wilhelminische Berlin. Zwischen Jahrhundertwende und Novemberrevolution, Berlin 1988, pp. 855–876; Schnall, Uwe: Staat und Seekartographie im wilhelminischen Deutschland, in: Lindgren, Uta (ed.): Kartographie und Staat. Algorismus 3 (1990), p. 61.

[8] Beiträge: p. 131 ff.

[9] Böhm: Überseehandel, p. 173 ff., 181.

[10] See also Güth, Rolf: Von Revolution zu Revolution. Entwicklungen und Führungsprobleme der Deutschen Marine 1848–1918, Herford 1978. p. 97; Führer durch das Museum für Meereskunde in Berlin, Berlin, 1907, p. 3.

[11] Witt: Reichsfinanzen, pp. 148–151.

[12] Westphal: Deutsche Kolonien, pp. 206–207.

[13] Witt, Peter-Christian: Die Finanzpolitik des Deutschen Reiches von 1903 bis 1913. Eine Studie zur Innenpolitik des Wilhelminischen Deutschland, Lübeck 1970, pp. 380–381; Witt: Reichsfinanzen, p. 146, pp. 151–153.

The empire's debts between 1898 and 1903 grew by nearly 627 billion marks to 2815 billion marks and were the reason for the German Empire's permanent financial crisis, lasting until World War I. After the elections on June 16, 1903, any previously common overdrafts or unbudgeted costs of the imperial departments were to be limited and the finances of the empire lastingly consolidated in order to finance the Reich's fleet-building activities on a long-term basis. These processes were initially very conducive to planning Germany's South Polar expedition; however, they produced the opposite results while their execution was still in progress. In other words, the South Polar expedition started in a time when copious funds could be generated, but it then suffered extremely from massive budget cuts in the years that followed.

The Road to Germany's First South Polar Expedition

On July 24, 1865, the First Convention of the Masters and Friends of Geography [Erste Versammlung Deutscher Meister und Freunde der Erdkunde] took place in Frankfurt and the group committed itself to the subject of polar research. Here, Georg von Neumayer appealed for the dispatch of a German South Polar expedition.[14]

However, priority was given to exploration of the closer, unknown Arctic, a proposition especially advocated by the cartographer August Petermann.[15] In 1868, the first German North Polar expedition explored the seas between the eastern part of Greenland and Spitsbergen. The second expedition (1868–1870) set out on two ships to explore the unknown eastern coast of Greenland as far north as possible. However, the two ships lost sight of each other after only a few weeks. The *Hansa* was crushed in the ice, and the crew members were forced to overwinter on a drifting ice floe before they finally managed to reach the Moravian mission station of Friedrichsthal on the southern tip of Greenland in their rescue boats. However, the second expedition group sailing on the *Germania* discovered Kaiser Franz Joseph Fjord and charted the coastline up to 77°N.

Neumayer's efforts in support of a German Antarctic expedition sparked general interest only when he drew a connection between South Polar research and measurement of the transit of Venus.[16] In 1874, astronomers wanted to observe the transit of Venus in front of the sun in order to determine the distance between the earth

---

[14] Kretzer, Hans-Jochen: Windrose und Südpol, Leben und Werk des großen Pfälzer Wissenschaftlers Georg von Neumayer, Bad Dürkheim 1984, p. 20; Neumayer, Georg von: Auf zum Südpol! 45 Jahre Wirkens zur Förderung der Erforschung der Südpolarregion 1855–1900, Berlin 1901, pp. 33–51.

[15] Krause, Reinhard A.: Hintergründe der deutschen Polarforschung. Von den Anfängen bis heute, in: Deutsches Schiffahrtsarchiv 16 (1993), pp. 13–30.

[16] Lüdecke, Cornelia: Die Routenfestlegung der ersten deutschen Südpolarexpedition durch Georg von Neumayer und ihre Auswirkung, in: Polarforschung 59 (1989), pp. 103–111.

Fig. 1 Georg von Neumayer, the hydrographer at the Admiralty, photographed in 1872. (Bundesamt für Seeschifffahrt und Hydrographie, Hamburg)

and the sun. As the Venus transit would be best seen in the Southern Hemisphere, Neumayer proposed the Kerguelen Islands in the Indian Ocean as a suitable vantage point, bearing in mind that an exploratory expedition could be easily sent from there to the unchartered region surrounding the South Pole.[17] Appointed to the post of hydrographer at the Admiralty in Berlin in the meantime, Neumayer succeeded in organizing a circumnavigation of the world by the SMS (Seiner Majestät Schiff [His Majesty's Ship]) corvette *Gazelle*, so that the transit of Venus could be observed on the Kerguelen Islands on December 9, 1874 (Fig. 1).[18] Neumayer, who became the director of the German Naval Observatory [Deutsche Seewarte] in Hamburg by 1876, subsequently advocated the route via the Kerguelen Islands as the best possible passage southward because the ocean currents there promised unimpeded progress.[19]

Independent of Neumayer's aspirations, Eduard Dallmann conducted a commercial whaling expedition on the steam bark *Grönland* to the Antarctic Peninsula in the southern summer of 1873–1874, an operation that also resulted in the discovery of Bismarck Strait, Neumayer Channel, and Kaiser Wilhelm Islands.[20]

More than ten years later, the Conferences of German Geographers [Deutsche Geographentage], which were held from 1885 to 1905 under Neumayer's

---

[17] Neumayer: Südpol, pp. 44–67.

[18] Kretzer: Windrose, p. 22; Lüdecke: Routenfestlegung, pp. 105–106; Headland, Robert Keith: A Chronology of Antarctic Exploration, London 2009, p. 206.

[19] Neumayer: Südpol, pp. 69–138, 347–350, 439–441.

[20] Headland: Chronologie, p. 203; Krause, Reinhard A./Ursula Rack (ed.): Schiffstagebuch der Steam-Bark Groenland geführt auf einer Fangreise in die Antarktis im Jahre 1873/1874 unter der Leitung von Capitain Ed. Dallmann, in: Berichte zur Polar und Meeresforschung 530 (2006).

presidency, provided an adequate forum for Germany's future South Polar research.[21] When the 11th Conference of German Geographers took place in Bremen on April 17–19, 1895, one session dedicated three lectures to research of the South Polar region.[22] Toward the end of the conference, the launch of a scientific South Polar expedition was motioned and unanimously accepted.[23] On the same day, the German Commission for South Polar Research [Deutsche Kommission für Südpolarforschung] was founded, with Neumayer as its president, Erich von Drygalski (subsequently the leader of the expedition), and 23 other members.[24] In their second session, an expedition plan was designed, which was to serve as the basis for discussion at the Sixth International Geographical Congress in London, held from July 26 to August 3, 1895, under the direction of Sir Clements Markham, the president of the Royal Geographical Society in London.[25]

Markham, campaigning for the dispatch of a British expedition to Antarctica, played a role similar to that of Neumayer in Germany.[26] At the congress in London, Neumayer presented his plan in which two ships were to make way southward on the Kerguelen route.[27] At the end, he referred to the advantages of an international collaboration in accordance with the ideal of the International Polar Year 1882–1883, when ten nations had, for a duration of 13 months, established 12 scientific stations for measurement of meteorological and magnetic data around the frozen Arctic Ocean. Finally, Neumayer's suggestions fell on fertile soil and Markham's hopes were also fulfilled. As a matter of course, Britain had to show its colors now, for it could not leave the exploration of the unknown South Polar region to other nations, let alone its political rival, Germany. Consequently, the congress culminated in a resolution that represented a millennium achievement, so to speak:

> The Sixth International Geographical Congress held in London 1895 regards the exploration of the Antarctic regions to be the most significant of problems yet to be solved and recommends, considering the benefits prospectively resulting thereof for all disciplines, that the various scientific societies all over the world should strive in ways they deem most effective to see that this task is accomplished before the close of the nineteenth century.[28]

---

[21] Meynen, Emil: Deutscher Geographentag 1881–1963. Gesamtinhaltsverzeichnis der Verhandlungen, Wiesbaden, 1965, p. VIII, 51.

[22] Drygalski, Erich von: Die Südpolarforschung und die Probleme des Eises, in: Verhandlungen des 11. Deutschen Geographen-Tages in Bremen im Jahr 1895, Berlin 1896, pp. 18–30.

[23] Friederichsen, Ludwig: Der sechste Internationale Geographen-Kongreß in London 26. Juli–3. August 1895. Mitteilungen der Geographischen Gesellschaft Hamburg 1895, p. 5.

[24] Drygalski, Erich von: Zum Kontinent des eisigen Südens, Berlin 1904, pp. 2–3; Lüdecke, Cornelia: Die deutsche Polarforschung seit der Jahrhundertwende und der Einfluß Erich von Drygalskis. Dissertation. Berichte zur Polarforschung 158 (1995), p. 133.

[25] Neumayer, Georg: Thätigkeitsbericht der Deutschen Kommission für die Südpolar-Forschung, in: Kollm, Georg (ed.): Verhandlungen des 12. Deutschen Geographen-Tages in Jena im Jahr 1897, Berlin 1897, p. 17; see also Friederichsen: Der sechste Internationale Geographen-Kongreß.

[26] Markham, Clements: Antarctic Obsession. A personal narrative of the origins of the British national Antarctic expedition 1901–1904 (posthumous), edited by Clive Holland, Alburgh 1986.

[27] Neumayer: Südpol, pp. 367–445.

[28] Friederichsen: Der sechste Internationale Geographen-Kongreß, p. 6.

In Germany, opinions collided in the subsequent sessions of the South Polar Commission because Drygalski, for practical reasons, had planned to use only one ship in order to get a chance to conduct the expedition, considering the tight financial situation the empire was in,[29] whereas Neumayer still demanded two ships for safety reasons.[30] On this point, Neumayer's influence on the South Polar Commission was obstructive, as it delayed concrete preparations. Neumayer just could not concede that his plan, cherished for decades, was now to be carried out contrary to his ideas by someone who was nearly 40 years his junior.

In August 1896, news made the headlines that the Norwegian Fridtjof Nansen, who had set sail on the *Fram* to the North Pole in 1893, had returned home safe and sound.[31] As planned, the *Fram* had been frozen in pack ice and drifted northward, but not far enough, for which reason Nansen and Hjalmar Johansen abandoned the ship in order to reach the pole by sledge and kayak. However, since huge ice floes were in permanent motion and failed to drift north, they had to turn around at 86°04′N and spend the winter in Franz Joseph Land, where they were finally rescued by the Briton Frederick Jackson. In the meantime, the *Fram* had come free of the ice again and succeeded in returning to Norway.

The Belgian expedition (1897–1899) led by Adrien de Gerlache de Gomery was, by now, the first to follow the call of the International Geographical Congress and explored the western side of the Antarctic Peninsula where, captured in the ice, it had to overwinter.[32]

As preparations for the German South Polar expedition came to a halt, Drygalski worked increasingly on publishing the results of his Greenland expedition (see his biography below) in order to submit them to the Friedrich Wilhelm University in Berlin as his (postdoctoral) habilitation thesis (Fig. 2).[33] Early in 1898, Drygalski completed the evaluation of his Greenland data. Shortly afterward, he received his lecturer qualification and three days later was appointed as the expedition leader in the last session of the South Polar Commission.[34]

---

[29] Drygalski: Zum Kontinent, pp. 3, 12–13.

[30] Neumayer: Südpol, pp. 349, 441.

[31] Nansen, Fridtjof: In Nacht und Eis. Die norwegische Polarexpedition 1893–1896. Reprint, Wiesbaden, 2011.

[32] Headland: Chronology, p. 229.

[33] Neumayer: Südpol, pp. 461–483; Drygalski, Erich von: (1948), Unpublizierte Autobiographie, private possession, Mörder, Feldkirchen-Westerham, pp. 32–34; Drygalski, Erich von: Aus dem nachgelassenen Lebensrückblick, in: Mitteilungen der Geographischen Gesellschaft in München 75 (1990), pp. 119–141, 72.

[34] Neumayer, Georg von: Zweiter Thätigkeitsbericht der Deutschen Kommission für die Südpolar-Forschung, in: Kollm, Georg (ed.): Verhandlungen des 13. Deutschen Geographen-Tages zu Breslau am 28., 29. und 30. Mai 1901, Berlin 1901, pp. 4–5. His long-postponed appointment was surely due to the circumstance that such a significant prestige project was not to be handed to a young "PhD" who was just 33 years old; instead, it should at least go to a person who had been scientifically acknowledged by completion of a habilitation thesis.

Fig. 2 Erich von Drygalski's dog sledge journey to survey operations in Greenland in 1893. (Leibniz-Institut für Länderkunde, Leipzig, Drygalski estate)

Erich von Drygalski

Erich von Drygalski was born in Königsberg (now known as Kaliningrad (Russia)) on February 9, 1865, as the middle son of five sons of the director of the renowned Kneiphof Gymnasium.[35]

In the winter semester of 1882–1883, he began studying physics, mathematics, and geography at the University of Königsberg. Other places of study were Bonn and Leipzig. Drygalski first encountered glaciers when he walked across the Alps for two months during the summer semester of 1884. Toward the end of his education, he went to Berlin in the winter semester of 1886–1887 to study with the renowned explorer Ferdinand Freiherr von Richthofen, and he earned his doctorate with a study in which he described the deformation of the globe by the weight of its ice cover during the glacials.[36]

(continued)

---

[35] Drygalski: Autobiographie, pp. 119–141.

[36] Drygalski, Erich von: Die Geoid-Deformation der Kontinente zur Eiszeit und ihr Zusammenhang mit den Wärmeschwankungen in der Erdrinde, Dissertation, in: Zeitschrift der Gesellschaft für Erdkunde zu Berlin 22 (1887), pp. 168–280.

However, he still required the essential basics to mathematically describe the physical conditions of the ice's movement during the glacial period. For this reason, Drygalski wanted to explore the movement of glaciers in nature first and then create a model derived from the measured data.[37] In the summer of 1891, he went on an exploratory trip to western Greenland, financed by the Berlin Geographical Society. The main expedition soon followed during 1892–1893.[38] The biologist Ernst Vanhöffen (a friend from his schooldays in Königsberg) and the meteorologist Hermann Stade accompanied him. A family of Greenlanders settled next to their overwintering station and, together with other locals, helped them with their excursions and surveys. Thus, they not only learned how to use dog sledges but also became acquainted with kayaking and other things necessary to survive in snow and ice.

On the basis of the two data volumes from this expedition, Drygalski submitted his habilitation thesis in Berlin 1898 and received his *venia legendi* ["authorization to read"] in geography and geophysics.[39] Only three days later, he was appointed as the leader of the South Polar expedition that had been planned long before. In the same year, the organization and management of the Physical–Geographical Department of the Marine Science Institute and Museum, recently founded by Richthofen, was also conferred on him (Fig. 3).[40]

From 1901 to 1903, Drygalski led the first German South Polar Expedition, which discovered Kaiser Wilhelm II Land in the Indian Ocean, close to the Antarctic Circle. By 1931, he had published 20 volumes and two atlases based on the results of the expedition.[41] When Drygalski's mentor, Richthofen, died suddenly in 1905, he took over his representation at both the university

(continued)

---

[37] Drygalski, Erich von: Über Bewegungen der Kontinente zur Eiszeit und ihren Zusammenhang mit den Wärmeschwankungen in der Erdrinde, in: Verhandlungen des 8. Deutschen Geographentages zu Berlin, Berlin (1889), pp. 162–180; Drygalski: Autobiographie, p. 52.

[38] Drygalski, Erich von: Grönland-Expedition der Gesellschaft für Erdkunde zu Berlin 1891 bis 1893, Vol. 1, Berlin 1897, pp. X–XI, 13–19; Lüdecke, Cornelia, Vor 100 Jahren: Grönlandexpedition der Gesellschaft für Erdkunde zu Berlin (1891, 1892–1893) unter der Leitung Erich von Drygalskis, in: Polarforschung 60 (1990), pp. 219–229; Lüdecke, Cornelia (ed.): Verborgene Eiswelten. Erich von Drygalskis Bericht über seine Grönlandexpeditionen 1891, 1892–1893, Munich, 2015.

[39] Drygalski: Autobiographie, p. 72.

[40] Ibidem, p. 79; Lüdecke, Cornelia: Erich von Drygalski und der Aufbau des Instituts und Museums für Meereskunde, in: Historisch-Meereskundliches Jahrbuch 4 (1997), pp. 19–36.

[41] Fels, Edwin: Erich Dagobert v. Drygalski, Neue Deutsche Biographie, Berlin (1959), pp. 143–144; see also Lüdecke, Cornelia/Heinz-Peter Brogiato/Ingrid Hönsch: Universitas Antarctica. 100 Jahre deutsche Südpolarexpedition 1901–1903 unter der Leitung Erich von Drygalskis, Leipzig 2001. The Drygalski estate, especially the material related to the South Polar Expedition, is archived at the Leibniz-Institut für Länderkunde in Leipzig.

Fig. 3 Erich von Drygalski, photographed in about 1901. (Gazert, Volker (Private possession))

the Marine Science Institute.[42] Finally, in 1906, he was offered the new chair in geography at the Ludwig Maximilians University in Munich.[43] Alongside this, he was the director of the Geographical Society of Munich from 1907 until his retirement in 1935.[44] In 1907, he married his cousin Clara Wallach, with whom he had four daughters.[45]

On account of his comprehensive polar experiences, Drygalski took part in the German Arctic Zeppelin Expedition in the summer of 1910, a study trip to explore the technical conditions for future zeppelin flights in the Arctic.[46]

After World War II, he directed the abandoned Geographical Institute of the University in Munich from the summer semester of 1947 until the winter semester of 1947–1948 and resumed giving lectures.[47]

On January 10, 1949, Erich Drygalski died in Munich and was buried in the cemetery at Partenkirchen.

---

[42] Drygalski: Autobiographie, p. 79, 108, 110; Lüdecke: Erich von Drygalski und der Aufbau.

[43] Fels: Drygalski, p. 143.

[44] Louis, Herbert: Die Geographische Gesellschaft München, Rückblick im hundertsten Jahre ihres Bestehens, in: Mitteilungen der Geographischen Gesellschaft in München 54 (1969), pp. 10–12.

[45] Fels: Drygalski, p. 143.

[46] Miethe, Adolf/Hugo Hergesell (ed.): Mit Zeppelin nach Spitzbergen, Berlin (1911); Meinardus, Wilhelm: Erich von Drygalski †, in: Petermanns Geographische Mitteilungen 93 (1949), p. 179.

[47] Personen und Vorlesungsverzeichnis für das Sommersemester 1947, Universität München, 1947, p. 54; Personen und Vorlesungsverzeichnis für das Wintersemester 1947/48, Universität München, 1947, p. 70.

Preparations for the First German South Polar Expedition

After Georg von Neumayer had successively retreated from further organization of the expedition and entrusted those responsibilities to Drygalski, the latter was able to put his own ideas into effect. In his preliminary "Plan of a German Expedition to the South Polar Region," dated February 22, 1898, Drygalski outlined the assignments of the expedition. Among them were oceanographic measurements; collection of plankton; magnetic and meteorological measurements both on board and on land; geological, zoological, and botanical collections; astronomical and geodetic measurements; geographic explorations on land and sea; and drift ice and land ice studies.[48] The magnetic and meteorological measurements were to chronicle the daily variation of the earth's magnetic field and its instabilities, as well as the climate of the little-known high southern latitudes.[49] Based on the new science of bacteriology (which had recently been founded and established by Robert Koch), bacteriological studies were also planned to take place in the course of the expedition and to be carried out by Hans Gazert, the scientifically interested physician for the expedition.[50]

For his expedition, Drygalski chose a form that is still valid today:

– Shipboard measurements on the outward-bound voyage in summer
– Station measurements during overwintering in Antarctica
– Exploration trips into the surroundings of the station in spring and summer
– Shipboard measurements on the way back in the southern autumn[51]

Drygalski's starting point for the geographic exploration of the unknown South Polar region was a fixed station from which he wanted to launch dog sledge expeditions. In his first expedition plan, he had envisaged using sledges to approach "the earth's pole, [and] in the next southern autumn follow the discovered coastline toward the magnetic pole."[52] In the subsequent versions of his plans, he still continued to mention the option of searching for the geographic and magnetic poles, although he strictly declined any one-sided striving for the pole, as his goal was of a purely scientific nature.

Early in April 1898, Drygalski negotiated for hours at the German Imperial Naval Office with the Chief of the Nautical Department, Graf Friedrich von Baudissin, and the secretary of state, Rear Admiral Tirpitz, about the nautical

---

[48] Drygalski, Erich von: Plan einer Deutschen Expedition in das Südpolargebiet, in: 17. Jahresbericht der Geographischen Gesellschaft München (1898), pp. 39–40.

[49] Bidlingmaier, Friedrich: Die erdmagnetisch-meteorologischen Arbeiten und Ausrüstungsgegenstände der deutschen Südpolar-Expedition und die Vorschläge für die internationale Kooperation während der Zeit der Südpolar-Forschung 1901–1903, in: Petermanns Geographische Mitteilungen 47 (1901), p. 153.

[50] Gazert, Hans: Bakteriologische Aufgaben der deutschen Südpolar-Expedition, in: Petermanns Geographische Mitteilungen 47 (1901), p. 153.

[51] See also Drygalski: Plan.

[52] Drygalski: Plan, p. 38.

issues involved in the expedition. A day later, he published a short draft describing the expedition ship.[53] Drygalski held the opinion that the expedition was a matter for the Department of the Navy and, as such, should be financed from the naval budget, and not from that of the Ministry of the Interior. "With it, one would be dealing only with a lower regulatory authority not as competently informed."[54]

Apart from the navy, which still had to be convinced to render its financial support, the Geophysical Commission of the Cartel Academies and Learned Societies of Gottingen [Geophysikalische Kommission der kartellierten Akademien und gelehrten Gesellschaften zu Göttingen]—which had formed from the Academies of Berlin, Munich, and Vienna and the Scientific Societies of Leipzig and Gottingen in order to support greater research projects—most warmly acknowledged Drygalski's expedition plan and particularly recommended its collaboration in geophysical questions that "promise making progress by way of corresponding observations made in remote places, especially in the Arctic and Antarctica."[55]

On July 20, 1898, the German Commission for South Polar Research, bypassing official channels, addressed an Immediate Submission directly to the kaiser in order to gain approval for 400,000 marks and 274,000 marks for 1899 and 1900, respectively, from the imperial budget for shipbuilding and for the Imperial Navy to take over leadership.[56] Time was running out, because an expedition (1898–1900) led by the Norwegian Carsten Borchgrevink had just embarked in order to overwinter for the first time on Antarctic soil, near Cape Adare in Victoria Land.

The Consultative Commission for the South Polar Expedition [Kommission für die Beratung einer Südpolarexpedition], instituted in accordance with the Immediate Submission, was convened early in November 1898. It was composed of Baudissin as its president, other representatives of the Imperial Naval Office and the Prussian Ministry of Ecclesiastical and Educational Affairs [Ministerium für geistliche und Unterrichtsangelegenheiten] (i.e., the Ministry of Cultural and Educational Affairs, represented by Friedrich Schmitt-Ott), and the expedition leader, and was supposed to discuss the building of the ship with regard to the measurement assignments, the duration of the expedition, and the costs.[57] Other representatives of the Ministry of

---

[53] Drygalski: April 5, 1898, Brief an Hans Meyer, private possession Wolfgang Kerler, Söcking; Oberhummer, Eugen: Die Deutsche Südpolarexpedition. Bericht über die vorbereitenden Schritte und die Versammlung in München am 13. Mai 1898, in: 17. Jahresbericht der Geographischen Gesellschaft in München 1898, pp. 40–42.

[54] Drygalski: April 5, 1899.

[55] Oberhummer: Bericht, p. 47; Baschin, Otto: Die Südpolar-Expedition, in: Zeitschrift der Gesellschaft für Erdkunde zu Berlin 36 (1901), p. 172.

[56] Kollm, Georg: Verhandlungen des 13. Deutschen Geographentages zu Breslau am 28., 29. und 30. Mai 1901, Berlin 1901, pp. 13–23.

[57] Kommission für die Beratung einer Südpolarexpedition: September 5, 1898, Sitzungsprotokoll, Leibniz-Institut für Länderkunde, Leipzig, Box 61, Inventory Number 4753, Serial Number 1.

the Interior and the German Hydrographic Office were invited to attend later sessions.[58]

In the meantime, Tirpitz furnished an opinion in which he confirmed the scientific and practical significance of the expedition. Should the government take over management of the expedition, and hence responsibility for its success and safety, he recommended using two ships "as it can be ruled out from the start that a potential failure could result from a lack of resources."[59] Further statements he made, however, related realistically to using only one ship, as in this case "the enterprise would be characterized as a private one and as such be supported by the empire." The essential tasks could be carried out by one ship alone. However, he believed that the submitted cost estimate was too low, and he increased the required sum of 874,000 marks to 1 million marks for using one ship, to be distributed over a period of four accounting years. The greatest share was assigned to the ship and the equipment "which would fall into the possession of the empire and represent a value beyond the purposes of the expedition, may it be that they were either sold or used for similar other purposes." As far as the Immediate Submission was concerned, the minister and state secretary of the Imperial Ministry of the Interior—as well as Vice Chancellor Graf Arthur von Posadowsky-Wehner (Fig. 4), who was the state secretary of the Imperial Treasury Office [Reichsschatzamt] from 1893 to 1897[60]—stated in an expert report commissioned by the Imperial Treasury that "there would be no second scientific task that incited greater interest than an advancement to the poles of the earth. A success of Germany in South Polar research, which has stagnated during the last 60 years, would add to the glory of German entrepreneurship and German science and meaningfully herald the new century with a peaceful exploit of German science."[61]

As to the question of the ship, Posadowsky finally received an assurance from Neumayer, in a confidential meeting that took place in early January 1899, that an expedition with only one ship could provide sufficient scientific results (Fig. 5).[62]

Besides the negotiations with the ministries, attempts were made to gain approval for the expedition on other levels as well. At a joint evening event of the

---

[58] Kommission für die Beratung einer Südpolarexpedition: November 19, 1898, Sitzungsprotokoll, Leibniz-Institut für Länderkunde, Leipzig, Box 61, Inventory Number 4753, Serial Number 1.

[59] Tirpitz: October 4, 1898, Tirpitz, Gutachten zum Immediatgesuch, GStA Merseburg, Rep. 92, Althoff estate Abt. B, No. 24, Vol. 2, Bl. 19–22. The following quotations also originate from this expert report.

[60] http://de.wikipedia.org/wiki/Arthur_von_PosadowskyWehner

[61] Posadowsky-Wehner: November 2, 1898, Brief an Thielmann, GStA Merseburg, Rep. 92, Althoff estate, Abt. B, No. 24, Vol. 3, Bl. 113–114, 125–126.

[62] Reichsministerium des Innern, March 10, 1899, Aktennotiz, BArch Potsdam, 15.01 RMdI, No. 16133, Bl. 4–6.

Fig. 4 Graf Arthur von Posadowsky-Wehner, the vice chancellor, minister, and secretary of state at the Imperial Ministry of the Interior. (Drygalski, Erich von: Zum Kontinent des eisigen Südens, Berlin 1904, p. 81)

Geographical Society of Berlin and the German Colonial Society (Section Berlin-Charlottenburg), which took place in Kroll's Theater on January 16, 1899, an audience of about 1300 assembled to listen to lectures focused on the objectives of the South Polar expedition. Among those who attended were the president of the Reichstag, representatives of the foreign office, envoys of foreign powers, high officers, and civil servants of the state.[63] As early as one week later, one deputy of the Reichstag advocated an additional budget for the expedition, as the latter would be "an important political assignment of the empire and a honorary obligation which we take upon ourselves if we are not to miss out on the forthcoming collaboration of civilized nations to explore the last part of our globe."[64] To this, Posadowsky answered that the navy would help by supplying the expedition with equipment, that it should set out with one ship in 1901, and that the budget of

---

[63] Verhandlungen: Gemeinschaftliche Sitzung der Gesellschaft für Erdkunde zu Berlin und der Abteilung Berlin-Charlottenburg der Deutschen Kolonialgesellschaft am 16. Januar 1899, in: Verhandlungen der Gesellschaft für Erdkunde zu Berlin 26 (1899), pp. 58–87; see also Oberhummer, Eugen: Die Deutsche Südpolarexpedition. Zweiter Bericht der Geographischen Gesellschaft in München, in: 18. Jahresbericht der Geographischen Gesellschaft in München 1900, p. 100.

[64] Oberhummer: Zweiter Bericht, pp. 102–105. The quotation is on p. 104.

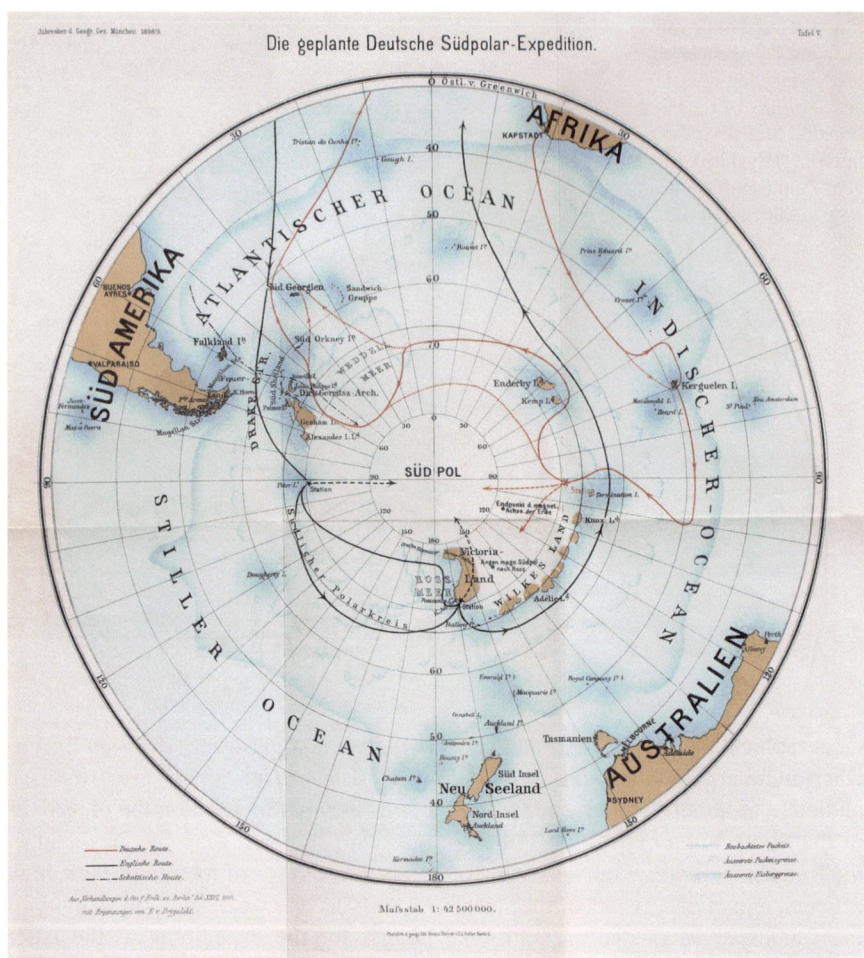

Fig. 5 The planned German South Polar expedition. The route of the German South Polar expedition to the magnetic South Pole and the geographic South Pole, along with a side trip to the Weddell Sea on the way back, as proposed by Neumayer in 1899. (Oberhummer, Eugen: Die Deutsche Südpolarexpedition. 18. Jahresbericht der Geographischen Gesellschaft in München 1900, table V)

1.1 million marks was to be spread over a period of five years. He also deemed it reasonable that the expedition should take place at the same time as the planned English expedition and a planned—albeit never accomplished—American expedition.

In its February session in 1899, the Budget Commission of the Reichstag took a unanimous resolution in which the government was requested to provide the required funds for the South Polar expedition.[65] Schmitt-Ott himself was able to

---

[65] Oberhummer: Zweiter Bericht, pp. 105–106.

draw 300,000 marks for equipment from the imperial budget.[66] By mid-March, the Imperial Treasury Office had raised no concerns about a state-financed South Polar expedition.[67] In April 1899, the kaiser finally gave his consent to include the costs of the South Polar expedition in the imperial budget.[68] Thus began the official support from the relevant authorities of the state.

The first official step, which indicated the actual securement of the endeavor to the public, was the appointment of a scientific council by the Ministry of the Interior. The "relationship between scientific control and nautical leadership" was discussed in the first council session, as was "the quality and number of scientific participants [and] the involvement of naval officers in solving research problems."[69] In a later session, it was stated that a branch station on the Kerguelen Islands in the southern Pacific Ocean was desired in order to obtain a station that was not influenced by Antarctica, for comparing meteorological and magnetic data measured in the south.[70] In the further course of the preparations, Tirpitz declared he was in favor of a captain coming from the merchant navy, as now it would no longer be an expedition of the navy and he would need all navy captains for the fleet.[71]

In late June 1899, the Consultative Commission for the South Polar Expedition was renamed the Commission for the German South Polar Expedition [Kommission für die deutsche Südpolarexpdition] and was still headed by the Imperial Naval Office. Because he had other obligations to meet, Baudissin handed over his presidency to Captain Ernst von Frantzius, whose primary responsibility was to take care of building the polar research vessel *Gauss*.[72] As a result of the conjoint preliminary work, the Reichstag received a "Memorandum Concerning the Equipment of a South Polar Expedition"—including a brief plan, map, and cost estimate—early in March 1900.[73]

---

[66] Schmidt-Ott, Friedrich: Erlebtes und Erstrebtes 1860–1950, Wiesbaden 1952. pp. 49–50.

[67] Thielmann: March 15 1899, Brief an das Reichsministerium des Innern, BArch Potsdam, 15.01 RMdI, No. 16133, Bl. 8.

[68] Oberhummer: Zweiter Bericht, pp. 116–118.

[69] Ibidem, p. 118.

[70] Wissenschaftlicher Beirat: November 24, 1899, Verhandlungen im Reichsministerium des Innern, Leibniz-Institut für Länderkunde, Leipzig, Box 61, No. 4754, Serial Number 2.

[71] Tirpitz: June 17, 1899, Brief an das Reichsministerium des Innern, BArch Potsdam, 15.01 RMdI, No. 16117, Bl. 117.

[72] Kommission für die deutsche Südpolarexpedition: June 27, 1899, Sitzungsprotokoll, Leibniz-Institut für Länderkunde, Leipzig, Box 61, Inventory Number 4753, Serial Number 1.

[73] Oberhummer: Zweiter Bericht, p. 130.

The *Gauss*
A special research vessel was built for the first South Polar expedition. It was made entirely of wood, with a hull made of highly durable timber and designed in a particular shape to withstand the pressure of sea ice (Fig. 6).[74] This not only made the vessel suitable for magnetic measurements, which at that time were impossible using iron-hulled ships, but also prevented the vessel from being crushed when it froze in, in which case it would be lifted upward, like Nansen's legendary *Fram*.

Fig. 6 The round hull of the *Gauss*, photographed in 1900. (Leibniz-Institut für Länderkunde, Leipzig, Drygalski estate)

(continued)

---

[74] Drygalski: Zum Kontinent, pp. 57–81.

Expedition ships before World War I, such as the *Fram* or the *Gauss*, were considered to be male, for which reason they were each referred to as "he" or "him."

The estimated total cost of about 1.2 million marks was to be distributed over five accounting years (Table 1).[75]

Table 1 Budgetary distribution for the first German South Polar expedition

Year	Budget (marks)
1899	200.000
1900	350.000
1901	510.000
1902	96.000
1903	54.000
Total	1.210.000

Technical Details of the Ship[76]

Three-masted topsail schooner with an auxiliary engine, built in 1901 at the Howaldt Shipbuilding Yard in Kiel (Figs. 7 and 8)

Length between perpendiculars: 46 meters

Breadth measured on the outer surface: 11.27 meters

Depth of hold: 6.3 meters

Engine: triple expansion engine (combined steam engine made of three units in serial order) with 325 horsepower

Speed loaded with 728 tons: 7 knots (approximately 13 kilometers per hour)

Real speed under steam: on average, 4–5 knots (approximately 7–9 kilometers per hour)

Special Equipment on the Ship

Wooden design, free of iron in the vicinity of the magnetic measurement instruments

Small winch for loads of up to 3.5 tons

Large fish winch for loads of up to 7.5 tons

Sixbee sounding machine [a machine for determining water depths]

Lucas sounding machine

Kite winder

Cable winch

Drinking water distillation apparatus

(continued)

---

[75] Oberhummer: Zweiter Bericht, p. 123.

[76] Drygalski: Zum Kontinent, pp. 57–81, 343–344; Kretschmer, Marine Oberbaurat: Die Südpolarexpedition, Berlin 1900. p. 12; Drygalski: October 9, 1900, Bestellung, Leibniz-Institut für Länderkunde, Leipzig, Box 64, Inventory Number 4767, Serial Number 1.

Fig. 7 The *Gauss*—the first German polar research vessel—photographed in 1901. (Oberhummer, Eugen: Die Deutsche Südpolarexpedition. 18. Jahresbericht der Geographischen Gesellschaft in München 1900. Table II)

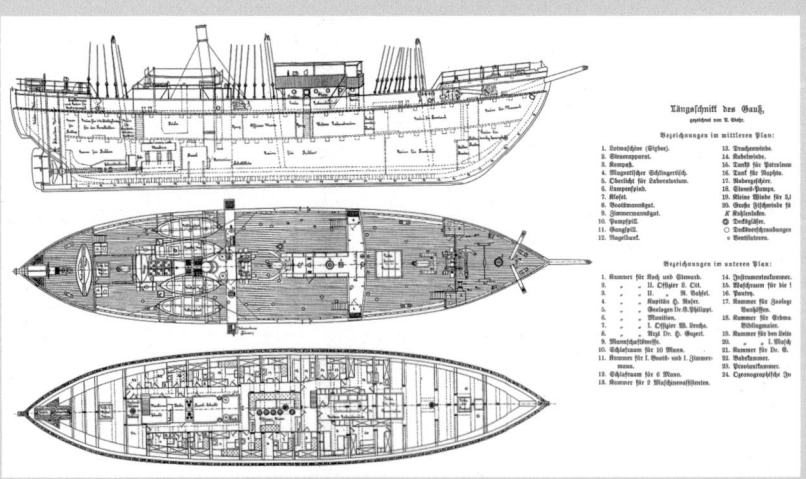

Fig. 8 A longitudinal section (*top*), the upper deck (*center*), and the living and working quarters (*bottom*) of the *Gauss*. (Drygalski, Erich von: Zum Kontinent des eisigen Südens, Berlin 1904, pp. 64–65)

(continued)

Fig. 9 The crew of the *Gauss*, photographed in 1902. *From left to right, third row*: Franz, Schwarz, Reimers, Heinrich, Michael, Lysell, and Johannsen; *second row*: Heinacker, Reuterskjöld, Marek, J. Müller, Dahler, Besenbrock, Klück, and Possin; *first row*: Bähr, L. Müller, Fisch, Stjernblad, Berglöf, Noack, and Björvik. (Leibniz-Institut für Länderkunde, Leipzig, Drygalski estate)

Wind engine
Electrical generator
Steam heater [never used]
Stones pump
Two captive balloons 300 cubic meters in volume
455 steel cylinders containing hydrogen
Crew: 27 men (Fig. 9)

Installation of a suspended table on the navigation bridge (used to measure the earth's magnetic field on a sea voyage) and skillful distribution of winches on an additional bridge allowed for several research operations to proceed simultaneously without entanglement of wires and cables carrying measurement devices lowered into the deep.[77]

The *Gauss* was the first German ship that had been built solely for research purposes and, to this end, it was provided with the latest technical equipment (Fig. 10).

(continued)

---

[77] Drygalski: Zum Kontinent, p. 14.

Fig. 10 The steam engine of the *Gauss*, photographed in Kiel in 1900. (Gazert, Volker (Private possession))

International Cooperation and Other Research Plans

When the Seventh International Geographical Congress was held in Berlin from September 28 to October 4, 1899, under the auspices of Drygalski's doctoral supervisor and supporter, Freiherr von Richthofen, an entire section was committed to polar research.[78] Here, Markham, who had decisively advocated for dispatch of a British Antarctic expedition under the direction of the Royal Navy officer Robert Falcon Scott, presented his ideas of a joint procedure in Antarctica.[79] He divided the presumed continent into four quadrants, assigning the Victoria and Ross Quadrants (between 90°E and 90°W) as an area of operation to Great Britain (because of her research tradition south of New Zealand) and the Weddell and Enderby Quadrants (south of the Atlantic, between 90°W and 90°E) to Germany (Fig. 11). Thus, the hypothetical continent was split among the sea powers into two territories of nearly identical sizes.

---

[78] Kollm, Georg: Verhandlungen des VII. Internationalen Geographen-Kongresses zu Berlin von 28. 9.–4. 10. 1899, Berlin 1901, p. 623 ff.; Fricker, Karl: Der VII. Internationale Geographenkongreß zu Berlin, Polarforschung, Geographische Zeitschrift 6 (1900), pp. 38–47.

[79] Markham, Clements: The Antarctic expeditions, in: Kollm: Verhandlungen des VII. Internationalen Geographen-Kongresses, pp. 623–630.

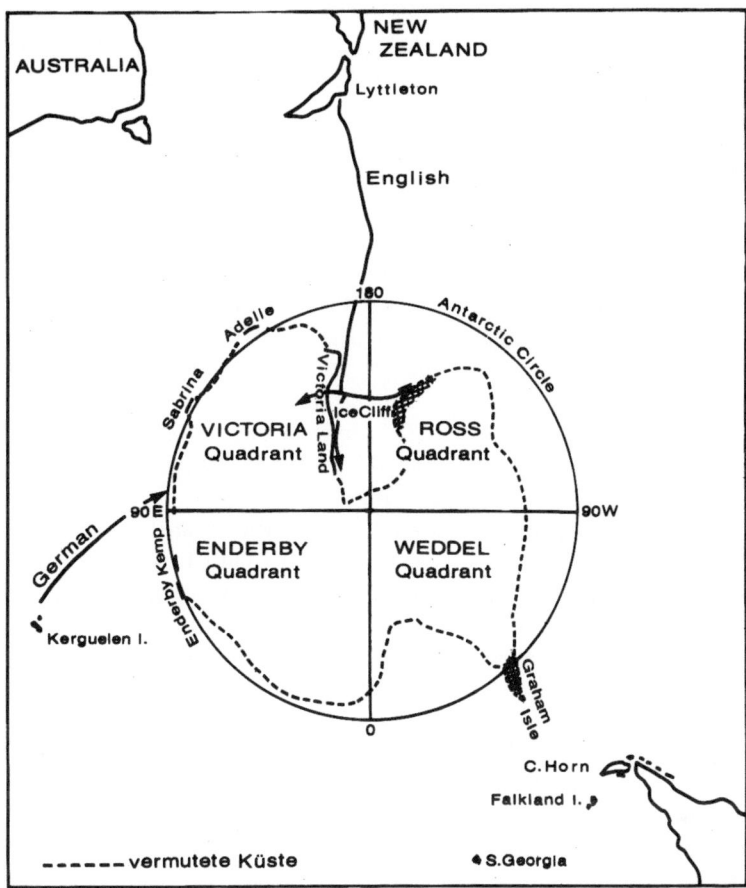

Fig. 11 Markham's sectoral division of Antarctica in 1899. (Lüdecke, Cornelia: Die erste deutsche Südpolar-Expedition und die Flottenpolitik unter Kaiser Wilhelm II. Historisch-meereskundliches Jahrbuch 1 (1992), p. 64)

Then Drygalski introduced his expedition plan according to which the unknown south was to be explored by a ship on the Kerguelen route.[80] At the end of his speech, he put a proposal to the vote in favor of establishment of international cooperation—which, after the model of the International Polar Year 1882–1883, was aimed at making simultaneous magnetic and meteorological observations—and this was passed unanimously.[81]

In Germany, the increased trade traffic with the colonies in the Southern Hemisphere prompted a demand for the kind of information needed to navigate ships

---

[80] Drygalski, Erich von: Plan und Aufgaben der Deutschen Südpolar-Expedition, in: Verhandlungen der Gesellschaft für Erdkunde zu Berlin 26 (1899), pp. 631–642.

[81] Drygalski: Plan und Aufgaben, pp. 641–642; Kollm: VII. Internationaler Geographen-Kongress, p. 85.

in these relatively unknown regions. Drygalski also placed his Antarctic expedition in the political context of Germany's fleet policy: "And the moment Germany gears up to develop her prestige at sea to an extent previously unknown, an expansion of naval knowledge where it is most lacking would be a national deed worth its price."[82]

Apart from the German and British expedition, the privately equipped Swedish expedition to Graham Land (now known as the Antarctic Peninsula) from 1901 to 1903, led by Otto Nordenskjöld on the *Antarctic*, was also part of the international cooperation, as was a Scottish expedition led by William Spears Bruce on the *Scotia*, which explored Coats Land (in the eastern Weddell Sea) from 1902 to 1904.[83]

To collect data for daily weather maps of the high southern latitudes, all involved stations on land and in the Antarctic, as well as all merchant and navy vessels south of 30°S, were to measure the air pressure, the temperature, the humidity, and the direction and strength of the wind, and make observations of clouds once a day at 0.00 PM Greenwich Mean Time (GMT) from October 1, 1901, to March 31, 1903.[84] To this end, booklets were handed out, which were to be collected afterward at the Office of the Geographical Congress in Berlin.

Drygalski had planned that meteorological parameters were to be observed every four hours every day during the voyage to Antarctica. Once in Antarctica, the operation of a station continuously measuring meteorological data all year round was to be established in order to obtain a measurement series of the Antarctic conditions (which were completely unknown) in all seasons.

To the network for absolute and variable geomagnetic measurements—which consisted of four overwintering stations in the South Polar region and their branch stations on the Kerguelen Islands (in the southern Indian Ocean), on Staten Island (in Argentina), on the Falkland Islands, and in New Zealand—the Germans added a geophysical observatory established by the Gottingen Academy of Sciences in German Samoa.[85] Here, too, single measurements were synchronized on an international level following the example of the International Polar Year (Fig. 12).

In March 1900, construction of the British polar vessel *Discovery* began in Dundee. On June 30 of the same year, Scott was at last officially appointed as the leader of the British expedition. One month later, the contract to build the first

---

[82] Drygalski: Aufgaben der Forschung, p. 133.

[83] Headland: Chronology, pp. 233–234, 236; Lüdecke, Cornelia: Scientific collaboration in Antarctica (1901–1903): a challenge in times of political rivalry, in: Polar Record 39 (2003), pp 25–48; Lüdecke, Cornelia: International Cooperation in Antarctica (1901–1904), in: Barr, Susan/Lüdecke, Cornelia (ed.): The History of the International Polar Years (IPYs), Berlin, Heidelberg (2010), pp. 127–134.

[84] Baschin: Südpolar-Expedition, pp. 53–57; Bidlingmaier: Erdmagnetisch-meteorologische Arbeiten.

[85] Luyken, Karl: Die erdmagnetischen Arbeiten auf der Kerguelen-Station, in: Kollm, Georg (ed.): Verhandlungen des 15. Deutschen Geographentages zu Danzig am 13., 14. und 15. Juni 1905, Berlin 1905, p. 58; Drygalski: Zum Kontinent, p. 24; Angenheister, Gustav G.: Geschichte des Samoa-Observatoriums von 1902 bis 1921, in: Birett, Herbert/Helbig, K./Kertz, Walter/Schmucker, U. (ed.): Zur Geschichte der Geophysik. Festschrift zur 50jährigen Wiederkehr der Gründung der Deutschen Geophysikalischen Gesellschaft, Berlin (1974), pp. 43–50.

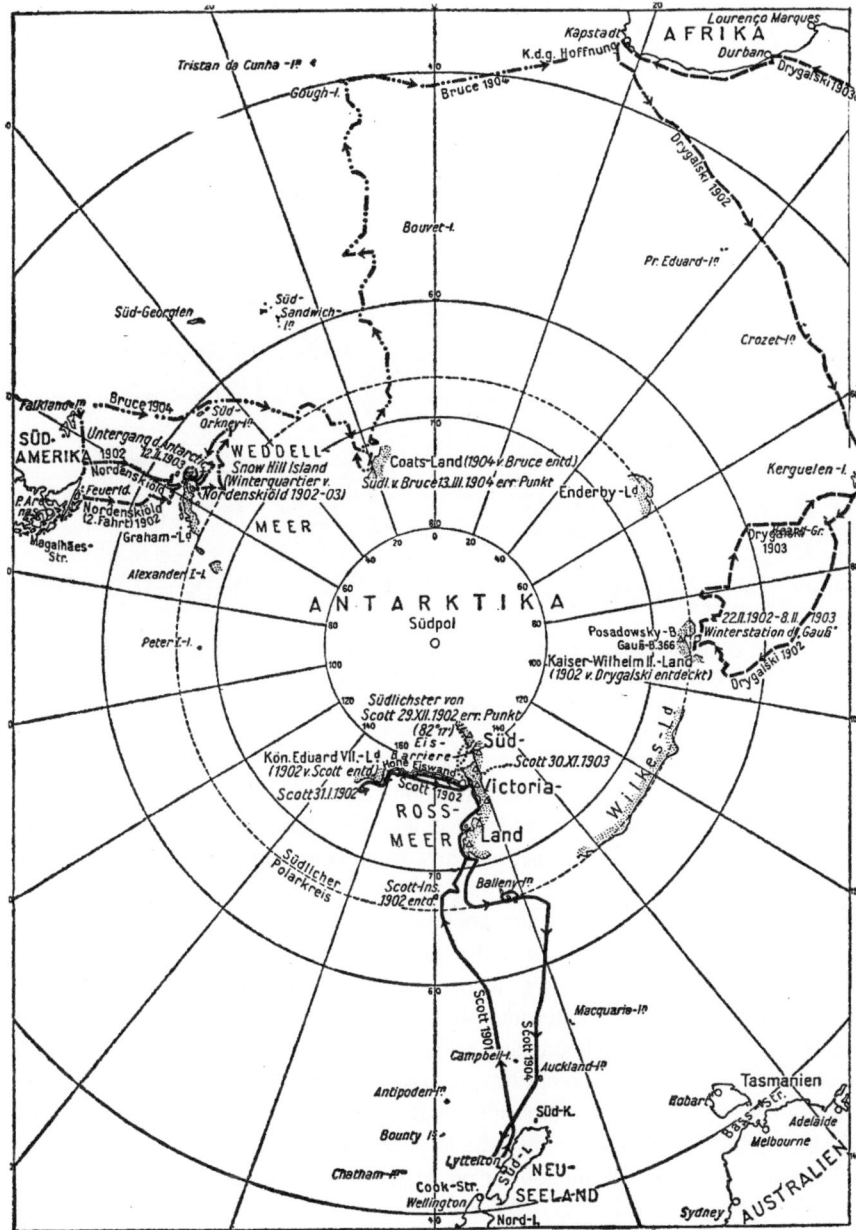

Fig. 12 The travel routes of the four cooperating Antarctic expeditions of Bruce, Drygalski, Scott, and Nordenskjöld (*shown in clockwise order*). (Förster, Hans Albert: Der Hohe Pol. Leipzig 1956, p. 203)

Fig. 13 The *Gauss* in frames at the Howaldt Shipbuilding Yard in Kiel, photographed in 1900. (Gazert, Volker (Private possession))

German polar research vessel was signed by the Ministry of the Interior and the Howaldt Shipbuilding Yard, and construction began in May 1900.[86] Later, the designated captain, Hans Ruser, and the chief machinist's mate [Obermaschinist], Albert Stehr, supervised the construction work in Kiel (Fig. 13).

As a result of the various events and reports about the planned expedition to the unknown, Drygalski received many applications from people wanting to participate in it. For example, a certain Wilhelm Filchner,[87] who would lead the second German South Polar expedition ten years later, applied late in 1900. Filchner had distinguished himself by an adventurous solitary ride across the Pamir mountains in the summer of 1900, but, as a lieutenant, he was unable to offer Drygalski any scientific qualification whatsoever, except for "willpower." Besides, all of the scientists had already been chosen at this point in time. Right from the beginning, Drygalski had had his friend Ernst Vanhöffen in mind to be the person responsible for the biological aspects of the expedition and for purchasing the polar equipment in Christiana (now known as Oslo (Norway)).[88]

---

[86] Reichsministerium des Innern: April 4, 1900, Vertrag mit den Howaldtswerken, BArch Potsdam, 15.01 RMdI, No. 16119, Bl. 27–30; Drygalski: Zum Kontinent, p. 14.

[87] Filchner: (end of 1900), Bewerbung, Institut für Länderkunde, Leipzig, Box 98, Inventory Number 6240, Serial Number 2.

[88] Drygalski: Zum Kontinent, pp. 17–18, 26–27.

The physician Hans Gazert had taken care of selecting the provisions and sports equipment for the expedition since April 1900. The geologist Emil Philippi had been intensively preparing for examining the deep-sea soil samples, while the physicist Friedrich Bidlingmaier had been trained at the Meteorological Geomagnetic Observatory in Potsdam since May. Drygalski would carry out the oceanographic and geographic studies himself.

After the *Gauss* had been launched on April 2, 1901, Otto Krümmel, an oceanographer in Kiel, turned to Drygalski in mid-April and explained to him, under a pledge of secrecy, his idea of a warm ocean current crossing the South Pole, for one should not be tempted to repeat the "experience Columbus had once . . . with the similarly hypothetical map drawn by Toscanelli."[89] The starting point was Nansen's drift across the Arctic Ocean toward the North Pole. With reference to the South Pole, Krümmel relied on "the Weddell Sea being temporarily free of ice" and on "the warm ocean current flowing southward from the Kerguelen Islands, hence towards the Antarctic Ocean," which had also been included in Neumayer's map (Fig. 14).[90]

Assuming that Enderby Land and Kemp Land were islands, Krümmel now boldly connected Wilkes and Victoria Land with the land sights pertaining to Graham and Alexander Land (now known as Alexander Island), and postulated the existence of a large ocean current, which extended from the Kerguelen Islands across the South Pole all the way to the Weddell Sea, "hence in analogy to the conditions prevailing in the Arctic Ocean. Then, a drift of the *Gauss* scheduled to exist between 90° and 100° eastern longitude was supposed to lead closely past the pole and into the Weddell Sea."[91] According to Krümmel, the *Gauss* would thus have a chance to drift poleward, just like Nansen's *Fram*. The remaining structure of the Antarctic continent was reminiscent of August Petermann's concept in that the northern coast of Greenland would be connected across the North Pole—with land sightings made north of the eastern Siberian coast (now known as Wrangel Island)—by means of a so-called elephant's trunk, whereby he deemed the remaining part north of Siberia to be free of ice.[92]

With the trial trip on May 24, 1901, and the acceptance of construction work associated with it, the work of the Commission for the South Polar expedition came

---

[89] Ibidem, p. 83, Krümmel: April 15, 1901a, Brief an Drygalski, Leibniz-Institut für Länderkunde, Leipzig, Box 61, Inventory Number 4755, Serial Number 3.

[90] Krümmel: April 15, 1901b, Die Trift des »Gauss« in den antarktischen Gewässern. Eine Prognose von O. Krümmel, Leibniz-Institut für Länderkunde, Leipzig, Box 61, Inventory Number 4755, Serial Number 3; see also Lüdecke, Cornelia: Ein Meeresstrom über dem Südpol? Vorstellungen von der Antarktis um die Jahrhundertwende, in: Historisch-meereskundliches Jahrbuch 3 (1995), pp. 35–50.

[91] Krümmel: April 15, 1901b.

[92] Tammiksaar, Erki/Natal'ja G. Suchova: August Petermann und seine Hypothesen über das Nordpolarmeer, in: Polarforschung 65 (1995), p. 134.

Fig. 14 Otto Krümmel's 1900 map of Antarctica, showing a presumed ocean current crossing over the South Pole. (Leibniz-Institut für Länderkunde, Leipzig, Drygalski estate: Box 61, No. 4755)

to an end.[93] The Commission was officially dissolved on the occasion of the 13th Conference of German Geographers in Breslau on May 28, 1901.[94] Drygalski devised a plan for the Ministry of the Interior to dispatch a second expedition should he not be able to send a message informing the ministry about his expedition prior to June 1, 1904.[95] After a preparatory period of more than six years, the first German South Polar expedition was now soon to begin (Fig. 15).

---

[93] Kommission für die deutsche Südpolarexpedition: May 24, 1901, Verhandlungsprotokoll an Bord der Gauss, BArch Potsdam, 15.01, RMdI, No. 16120, Bl. 198–201.

[94] Kollm: Verhandlungen des 13. Deutschen Geographentages, p. XVII.

[95] Drygalski: July 15, 1901, Brief an das Reichsministerium des Innern, GStA Merseburg, Rep. 76 Vc, Sekt.1, Title 11, Part VA, No. 7, Vol. I, Bl. 396–402.

Fig. 15 Scientists and officers on board the *Gauss* in 1901. *From left to right*, *back row*: Ott, Bidlingmaier, Werth, Gazert, and Philippi; *front row*: Vahsel, Vanhöffen, Drygalski, Ruser, and Lerche. (*Terra Marique* 1902, Issue 1)

Orders for the South Polar Expedition

Unlike the earlier North Polar expeditions, the South Polar expedition was not bound by any extensive instructions that had been decided from behind office desks. Instead, the decree signed by Kaiser Wilhelm II on July 18, 1901, was short and of a general nature: "The expedition shall leave Kiel in the month of August and make way to the Kerguelen Islands. There, a magnetic–meteorological station is to be built. Then the southbound voyage shall be continued. The Indian–Atlantic side of the South Polar region is to be considered as the research area. Should the continent be reached, a research station shall be established on [the] same, if possible, and be maintained whatever it takes for the duration of one year. The return voyage shall begin according to the expedition leader's orders in the spring of 1903 or at the latest in the spring of 1904."[96]

In this regard, the executive orders emphasized the "principle of freedom of movement for the executing personages."[97] The same decree also appointed Drygalski as the commander of the South Polar Expedition (Fig. 16). He had

---

[96] Drygalski: Zum Kontinent, pp. 52–53.
[97] Ibidem, p. 53.

Fig. 16 A 1901 greeting card from the German South Polar Expedition, showing the *Gauss* on the high seas and a flag inscription pointing "To the South Pole". (Gazert, Volker (Private possession))

recruited, with great circumspection, the nautical and small scientific research staff required to perform numerous tasks. He also attributed great importance to the crew assisting the scientists in their measurements while overwintering, so that they were permanently occupied and escaped boredom. For this reason, the seamen Georg Noack and Georg Wienke had been trained as taxidermists at the Natural Science Museum in Berlin and, as such, were able to support the biologists in their work.[98]

The Ministry of the Interior issued a special official directive encompassing 22 articles for the South Polar expedition, which primarily defined the relationship between the scientific leader of the expedition and its nautical leader.[99]

The relationship between the crew and captain were determined by the Regulations for Seamen [Seemannsordnung] and additionally adapted to the polar expedition by an official directive for the ship's crew.[100] As the expedition was financed by the government, the *Gauss* was entitled to sail under the imperial flag of the Ministry of the Interior [Reichsdienstflagge des Innenministeriums], which

---

[98] Ibidem, p. 44, 51.

[99] Reichsministerium des Innern: July 18, 1901a, Dienstanweisung für die »Deutsche Südpolar-Expedition«, Leibniz-Institut für Länderkunde, Leipzig, Box 70, Inventory Number 4870, Serial Number 6. A representation of the most important aspects is found in Drygalski (Zum Kontinent, pp. 53–56); see also the representation of all paragraphs in note form in Lüdecke, Cornelia: Die erste deutsche Südpolar-Expedition und die Flottenpolitik unter Kaiser Wilhelm II, in: Historisch-meereskundliches Jahrbuch 1 (1992), pp. 72–74.

[100] Reichsministerium des Innern: July 18, 1901b, Dienstanweisung für die »Deutsche Südpolar-Expedition« (Schiffsmannschaft), BArch Potsdam, 15.01 RMdI, No. 16121, Bl. 133–141, 231–233.

Fig. 17 An official service cap, bearing the emblem of the German South Polar Expedition (SPE). (Gazert, Volker (Private possession))

consisted of the flag of the German Empire (showing the colors black, white, and red) and the imperial crown in a circular white field at its center.[101]

The expedition also had its own special service emblem. To underscore the nautical character of the expedition, each scientist received a service cap bearing the emblem of the South Polar Expedition (SPE), which was composed of two crossed anchors and the seal of the Imperial Ministry of the Interior (Fig. 17). As he was the leader of the expedition, Drygalski was the official representative of the Ministry of the Interior, both outwardly and on board (Fig. 18).

Drygalski outranked even the captain, for the ship (including its personnel and equipment) was placed at his absolute disposal, provided that life and the ship itself were not at risk. The ship's crew was also under his command, as long as the captain's authority and duties on board were not affected. This most singular order surely contributed to the success of the expedition. In addition, it had been determined that Drygalski was to voluntarily lay down his functions and transfer the leadership of the expedition to his deputy in the event of a "pathological overstimulation of his nervous system . . . and after seeking the advice of a physician."[102] Furthermore, a "secret order" determined that Drygalski was entitled to suspend the captain from his duties should the latter "significantly endanger" the fulfillment of expedition assignments for health reasons or act "in private opposition to the orders of the leader."[103] The entire procedure was to be handled in confidence. Only if the captain refused to resign voluntarily was his suspension to be decided unanimously on the basis of a medical report issued by a committee consisting of the physician, Drygalski's deputy (Vanhöffen), and the officer next in rank to the captain. Fortunately, Drygalski did not have to make use of this secret order (Fig. 19).

---

[101] Reichsministerium des Innern: July 18, 1901a; Baschin: Südpolar-Expedition, p. 199.

[102] Posadowsky: July 27, 1901a, Brief an Drygalski, Leibniz-Institut für Länderkunde, Leipzig, Box 466, File d, Serial Number 98.

[103] Posadowsky-Wehner, Graf Arthur von: July 27, 1901b, Geheime Ordre, Leibniz-Institut für Länderkunde, Leipzig, Box 466, File d, Serial Number. 109.

Fig. 18 The official seal of the Imperial Ministry of the Interior for the German South Polar Expedition. (Cornelia Lüdecke (Private possession))

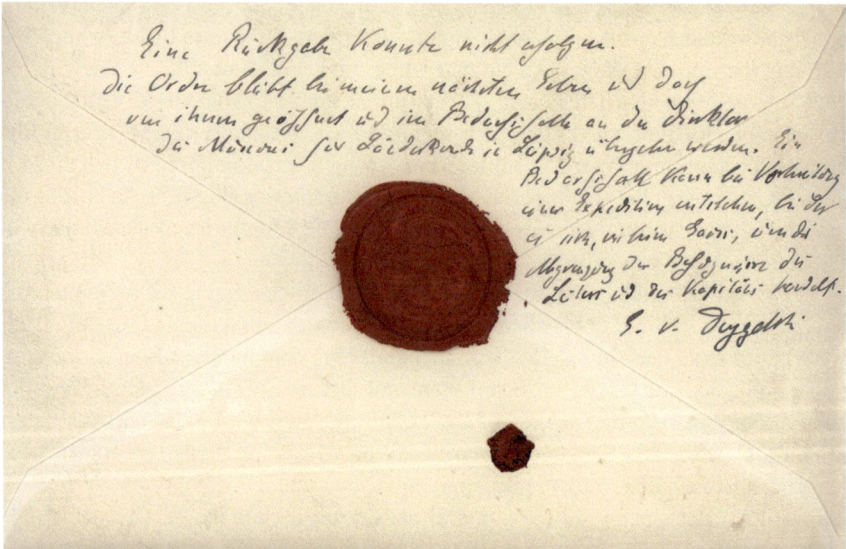

Fig. 19 The "secret order" that Drygalski forwarded unopened to the Museum of Regional Geography (now known as the Leibniz Institute for Regional Geography). (Leibniz-Institut für Länderkunde, Leipzig, Drygalski estate: Box 466d No. 109)

Over and above it stood the general duties of the participants in the expedition "to reach the goal common to all by mobilizing all forces. This required from everyone unreserved commitment, unconditional subordination, selflessness and loyal comradeship in all, even the most difficult situations."[104] In return, the Ministry of the Interior guaranteed all expedition members full pensions in cases of invalidity not compensated for by the Marine Accident Insurance Law.[105]

---

[104] Reichsministerium des Innern: July 18, 1901a: § 7.

[105] Reichsministerioum des Innern: July 27, 1901, Brief an Drygalski, Leibniz-Institut für Länderkunde, Leipzig, Box 467, File k, Serial Number 185.

During the whole expedition, Drygalski advocated the "principle of liberty [and] responsible and prompt decision-making on-site for everyone within his range of competence, but in service to the entire mission."[106] This principled stance benefited from the fact that only a very generalized framework directive had been issued for the expedition. It enabled experimentation with various research approaches and means in the unknown work fields of the Antarctic. The South Polar Expedition was therefore the last of its kind, being totally bound to Humboldtian ideals of research, i.e., examining everything from all sides as comprehensively as possible. Later expeditions became more and more specialized in their assignments.

The Financial Background During the Expedition
Right after the departure of the *Gauss*, the Imperial Treasury Office reported to the Ministry of the Interior that, according to the budget plan, the financial resources for the expedition would last only until early 1903. to be done if the expedition needed to be prolonged by a year.[107] The reply stated that the expedition members had been commissioned for a period of three years and that in normal circumstances the duration of the expedition would be two years, and three years only in the event of an emergency.[108] Thereupon, the Imperial Treasury Office made Posadowsky from the Ministry of the Interior personally aware that the budget for the expedition members' wages would not last until 1904.[109] Late in 1901, a third memorandum documented an increase in the expedition costs by 309,000 marks to a total of 1,509,000 marks[110]; thus—reaching beyond the budget of 1901—200,000 marks would have to be offset but could be compensated for.[111] A total of 109,000 marks remained for 1903. However, according to a fourth memorandum, this sum grew and ultimately resulted in a cost exceedance of 150,000 marks for the year 1903, meaning that the expedition budget was overdrawn by about 460,000 marks altogether.[112] Drygalski was not informed of this development.

---

[106] Drygalski: Zum Kontinent, pp. 51–52.

[107] Reichsschatzamt: August. 13, 1901, Brief an das Reichsministerium des Innern, BArch Potsdam, 15.01 RMdI, No. 16122, Bl. 85.

[108] Reichsministerium des Innern: August 24, 1901, Brief an das Auswärtige Amt, BArch Potsdam, 15.01 RMdI, No. 16122, Bl. 86–87.

[109] Thielmann: September 13, 1901, Brief an Posadowsky, BArch Potsdam, 15.01 RMdI, No. 16125, Bl. 2–3.

[110] Oberhummer, Eugen: Die Deutsche Südpolarexpedition. Dritter Bericht der Geographischen Gesellschaft in München, in: 19. Jahresbericht der Geographischen Gesellschaft in München 1901, pp. 120–121.

[111] Reichsschatzamt: September 27, 1901, Brief an die Reichshauptkasse, BArch Potsdam, 15.01 RMdI, No. 16136.

[112] Reichsministerium des Innern: October 14, 1902, Entwurf der vierten Denkschrift, BArch Potsdam, 15.01 RMdI, No. 16137, 3 Bl.

The Voyage to the South

Because Victoria, the Empress and Queen of Prussia, had died on August 5, 1901, at the age of 60 years, general mourning forbade holding a grand farewell party for the expedition.[113]

On August 11, 1901, the *Gauss* set sail with bursts of triple hurrahs from the surrounding battle ships. The passage led through Kaiser Wilhelm Canal (now known as Kiel Canal) to the Lower Elbe, where the voyage was interrupted close to the lighthouse of Neuwerk for three days for final loading of stores and getting the ship shipshape.

On board the ship were[114]:

> Approximately 20,000 kilograms of meat and sausage in about 80 varieties
> 13,000 kilograms of flour
> 10,000 kilograms of bread, hard bread, biscuits, cakes and zwieback in 20 varieties
> 700 kilograms of vegetables of all kinds
> 6000 kilograms of sugar
> 500 kilograms of butter and lard
> 4000 kilograms of condensed milk
> 2000 kilograms of soups in 37 varieties
> 2000 kilograms of jelly, jam, fruit juices, candies and dried fruit
> 1500 kilograms of cocoa
> 1000 kilograms of coffee
> 160 kilograms of tea, etc.

On August 15, the anchors were raised and the *Gauss* finally started on her southbound voyage.[115] For traditional reasons, the Kerguelen route favored by Neumayer was to be taken to get as far south as possible. There were some interesting oceanographic measurements to be made on the way across the Atlantic because the route was designed to pass between the routes of the *Challenger* and the *Gazelle*, and thus through unchartered sea.[116]

First of all, Drygalski was to cross the equator below 18°W in order to confirm the depth soundings of over 7000 meters previously made by the French battleship *Romanche* at 0°11′S and 18°15′W, as those findings had not yet been entered into the English and German sea maps.[117] The *Gauss* therefore deviated from her regular sailing route to Cape Town and unexpectedly encountered an area that was characterized by a light northeast trade wind. Here, only little progress could be made and the voyage had to be continued with steam power (Fig. 20).

---

[113] Drygalski: Zum Kontinent, pp. 84–85, 188.

[114] Bidlingmaier, Friedrich: Zum ewigen Eise des Südpolar-Kontinents, in: Deutsches Knabenbuch 18 (1904), pp. 52–53.

[115] Drygalski: Zum Kontinent, pp. 85–86.

[116] Oberkummer: Dritter Bericht, p. 130.

[117] Drygalski: Zum Kontinent, pp. 103, 108–109.

Fig. 20 The travel route of the *Gauss* from Kiel to Antarctica. (Drygalski, Erich von (ed.): Deutsche Südpolar-Expedition 1901–1903: Vol. 7, Table 3)

Fig. 21 Depth sounding in heavy seas. (Mörder, Thomas (Private possession))

The soundings of the Romanche Deep near the Mid-Atlantic Threshold (now known as the Mid-Atlantic Ridge) confirmed the exceptionally deep (7370 meter) depression in the Atlantic (Fig. 21).[118]

A haul produced an astounding discovery: "The emergence of a white spot is welcomed with rejoicing and soon the slender, white deep-sea net hangs alongside [the] ship after having been drawn through the water vertically from a depth of 4,000 meters. It [has] brought up a huge amount of peculiar deep-sea creatures with their adventurous black and red colors, illumination organs, stalked eyes and strange mouthparts. Many organs are torn: they are built for the tremendous water pressure in the deep and must burst due to the tremendous pressure decrease when they are hauled up. The zoologist comes with cans and bottles and rapidly retrieves his treasures before a crowd of curious onlookers." (Figs. 22, 23 and 24)[119]

On the next day (October 2), the traditional line-crossing ceremony took place. Even the leader of the expedition was not spared, for, after "soaping [of] one's face under ceremonial salutations and after its completion [proceeded] a backward fall into the vat."[120] The festive day came to an end with beer, cigars, and hilarious speeches (Fig. 25).

---

[118] Bidlingmaier: Zum ewigen Eise, p. 54.
[119] Ibidem, pp. 54–55.
[120] Drygalski: Zum Kontinent, p. 110.

Fig. 22 The zoologist Vanhöffen fishing for specimens in the surface waters of the Atlantic. (Leibniz-Institut für Länderkunde, Leipzig, Drygalski estate)

Fig. 23 The Sixbee sounding machine being used for taking soundings. (Leibniz-Institut für Länderkunde, Leipzig, Drygalski estate)

Fig. 24 Gazert drawing blood from a shark for biological analyses. (Leibniz-Institut für Länderkunde, Leipzig, Drygalski estate)

Fig. 25 "Neptune" and his entourage have come on board for the line-crossing ceremony. (Mörder, Thomas (Private possession))

When the ship was stationary again for soundings on October 30, Gazert shot the first albatross.[121]

As the measurements were tedious, the time lost could not be made up for.[122] To make matters worse, an alarm was sounded because a leak had occurred in the rear section of the ship and flooded the engine room to some extent, and so the steam pumps had to keep running day and night.[123] The leak also contributed to growing restlessness among the crew members. The expedition arrived in Cape Town only on November 23, 1901, not on October 20 as had been planned. There, martial law ruled the country because of the Second Boer War (1899–1902), but this did not have any significant impact on the expedition. First of all, the leak on the *Gauss* had to be fixed (Fig. 26).

In addition, various crew members who had proved to be unfit for the voyage had to be replaced. The expedition's physician made himself useful and his rounds in Cape Town. In addition, Drygalski, Vanhöffen, Gazert, and the biologist Emil Werth, who was intended to lead the station on the Kerguelen Islands,[124] climbed Table Mountain, guided by their fellow countryman Rudolf Marloth, who was the president of the local mountaineering association. There also were several invitations to banquets. Only after the repairs had been finished was the *Gauss* able to leave Cape Town on December 7, 1901.[125]

Shortly after the *Gauss* set sail, two blind passengers were discovered on board, for whom Drygalski found useful employment during the expedition. In this regard, it was later acknowledged that "it is one of the great achievements of our leader how he succeeded in occupying the life of each and every one, down to the last deckhand, with a responsible, independent duty. He has made everyone feel: 'I am the authority in my field of work, this gives my life value, and I am thus a significant part of the whole.'"[126]

In the meantime, the meteorologist Josef Enzensperger and the physicist Karl Luyken, along with the seaman Wienke, had brought part of the equipment first to Sydney on board the Lloyd steamship *Karlsruhe* and then—after taking over 30 dogs from Kamchatka and provisions and further equipment for the branch station—to the Kerguelen Islands on board the Lloyd steamer *Tanglin* (Fig. 27). There, in the second half of November, they eagerly awaited the arrival of the *Gauss*, which carried further provisions for the station.[127]

---

[121] Ibidem, p. 127.
[122] Oberhummer: Dritter Bericht, p. 115; Drygalski: Zum Kontinent, pp. 139–145.
[123] Drygalski: Zum Kontinent, pp. 133–134.
[124] Ibidem, pp. 36, 51.
[125] Ibidem, pp. 158–159.
[126] Bidlingmaier: Zum ewigen Eise, p. 66.
[127] Drygalski: Zum Kontinent, pp. 36–38, 51, 189; Enzensperger, Josef J.: Die deutsche Südpolarexpedition, in: Petermanns Geographische Mitteilungen 48 (1902), p. 70; Bidlingmaier: Zum ewigen Eise, pp. 56–57.

Fig. 26 The *Gauss*, docked in Cape Town between November and December 1901. White army tents can be seen *in the background*. (Gazert, Volker (Private possession))

Fig. 27 Karl Luyken (*third from the right*) and Joseph Enzensperger (*far right*) reached the Kerguelen Islands on the *Tanglin* in 1901. (Leibniz-Institut für Länderkunde, Leipzig, Drygalski estate)

Once the *Gauss* had left the African coast, the fair weather conditions alternated with rough seas and fog. When it cleared again on Christmas Day, the first icebergs suddenly emerged portside—a great sensation that far north (Fig. 28).[128]

The next stop was the Crozet Islands, where the expedition paid Possession Island a visit on Christmas Day.[129] Arrival on the Kerguelen Islands proceeded on January 2, 1902, a total of six weeks later than had been planned (Fig. 29).[130] Here, the expedition stayed until the end of the month in order to assist in the further establishment of the station (Figs. 30, 31 and 32), for the Chinese crew on the *Tanglin* had suffered greatly from beriberi and had grown much too weak to engage in physical activities. Two seamen died and were buried on land. What was behind this disease, induced by vitamin B_1 deficiency (then referred to as polyneuritis),

---

[128] Drygalski: Zum Kontinent, p. 166.
[129] Ibidem, pp. 167–173.
[130] Ibidem, pp. 157, 181.

Fig. 28 The first icebergs drawn in Gazert's watercolor sketchbook on February 12–13, 1902. (Gazert, Volker (Private possession))

Fig. 29 The *Gauss* in Observatory Bay on the Kerguelen Islands. (Cornelia Lüdecke (Private possession))

Fig. 30 Kerguelen Station. Station Mountain can be seen *in the background*. (Cornelia Lüdecke (Private possession))

Fig. 31 *From left to right*: Drygalski, Ruser, Bidlingmaier, Enzensperger, and Werth standing in front of Kerguelen Station. (Leibniz-Institut für Länderkunde, Leipzig, Drygalski estate)

Fig. 32 On the Kerguelen Islands. Variation House, in which short-term changes in the magnetic field were measured, can be seen *in the foreground*. The observatory, which was used for making absolute measurements of the magnetic field, is visible *in the background*. (Cornelia Lüdecke (Private possession))

was still unknown around 1900.[131] However, it was already known that Kerguelen cabbage (*Pringlea antiscorbutica*), growing in huge shrubs at that time, prevented vitamin C deficiency. During the layover, Drygalski decided on the final orders for the rescue expedition since its deadline had been brought forward by a year to June 1, 1903.[132]

On January 31, 1902, after loading of the previously deposited coal, additional provisions, and the dogs, the southbound voyage finally began. After a short stop on Heard Island and a close encounter with sea elephants, the first snow fell and the

---

[131] Gazert, Hans/Renner, Otto: Die Beriberifälle auf Kerguelen, in: Drygalski, Erich von (ed.): Deutsche Südpolar-Expedition 1901–1903 im Auftrage des Reichsamtes des Innern, Vol. 7, Issue 4, Berlin 1914, pp. 353–386.

[132] Drygalski: January 1, 1902, Brief an das Reichsministerium des Innern, GStA Merseburg, Rep. 76 Vc, Sect.1, Title 11, Part VA, No. 7, Vol. I, Bl. 410–417.

Fig. 33 A seal hunters' hut on Heard Island, photographed on February 3, 1902. (Mörder, Thomas (Private possession))

weather became quite changeable (Fig. 33).[133] In the storm, the *Gauss* danced on the waves and took on so much water that the deck crew at times stood chest-high in water. An attempt to calm the waves by deliberately pouring "oil on troubled waters"—a method commonly used to calm rough seas at that time—did not succeed, because the storm oil had already become too viscous and would no longer drip.

On February 8, the crew was able to observe the first polar lights the whole night through, which conveyed a notion of the beauty of the coming polar night.

Because of the advanced season, the *Gauss* encountered extended fields of drift ice as early as mid-February, which soon grew denser in the course of time. It began to snow, and a storm was approaching.

Below the log entry on February 21, 1902, Drygalski eventually noted that "We got the land!" as the crew sighted uniformly white contours on the horizon which, as they correctly inferred, was inland ice. They christened the mainland (which they assumed lay underneath) Kaiser Wilhelm II Land.[134]

There was great joy at first, but the next morning, the *Gauss* got stuck and could not be moved. There were plenty of discussions as to whether it would have been better to have taken another course. In a later account of the expedition, Drygalski wrote, "Never again did I feel what it means to take decisions as on that evening:

---

[133] Drygalski: Zum Kontinent, pp. 211–229.

[134] Ibidem, pp. 241–246.

there were no prior indications of it and therefore many opinions existed, not allowing any advice to be deduced. Finally, after due considerations one acts on one's own intuition, as fate is determined by the elements."[135] On February 22, the *Gauss* was frozen in the ice at a position of 66°2′S and 89°48′E, nearly 80 kilometers off the Antarctic mainland. This was much farther north than the station Scott had planned at 77°30′S and 167°42′E on Ross Island. Indeed, Neumayer had hoped that a deep trench would reach southward from here, as in the Ross Sea.

Overwintering in the Ice

When viewed from the outside, the South Polar Expedition appeared to be rather harmonious. But in his personal diaries, Drygalski wrote about problems on board, which he concealed in his subsequently published travel account. As early as on their voyage to Cape Town, he took notice of Captain Hans Ruser's unpleasant behavior and, now entrapped in the ice, the officers refused to follow his exaggerated orders.[136] Unfortunately, the excerpts from Drygalski's diary did not disclose any details about Ruser's behavior. At any rate, time and again, Drygalski had to bring balance to the situation and resolve conflicts.

When it became clear that the situation in the ice would not change, the men began preparing to overwinter (Fig. 34).[137] In particular, the officers were in high spirits now that the sea voyage under Captain Ruser had come to a temporary end and leadership on land was transferred to Drygalski.[138] The scientists who were making preparations for their observations had to contend with some difficulties. A field forge and a carpenter's workshop were set up on the ice, and "magazines were built in which the stockpiled treasures of the *Gauss* were [laid] out in neatly and practicable arrangement. The dogs got a long wood shed to protect them against the wind, the maternal and young animals a warm palace made of crates and cork tiles (Fig. 35)."[139]

Meteorological instruments were mounted close to the ship (Fig. 36).[140]

Friedrich Bidlingmaier, who originally had also been assigned to meteorology, had to give up this area of responsibility because the magnetic measurements demanded his full attention. Therefore, the physician Hans Gazert took over the supervision of meteorological duties and was joined by the geologist Emil Philippi

---

[135] Ibidem, p. 246.

[136] See also the diary excerpt in note form from Drygalski 1901–1903a, Stichwortartiger Auszug aus Drygalskis Südpolartagebüchern, private possession Gazert, Partenkirchen, and, in particular, the entry made on February 25, 1902.

[137] Drygalski: Zum Kontinent, pp. 254–264.

[138] Ibidem, pp. 266–267.

[139] Bidlingmaier: Zum ewigen Eise, p. 64.

[140] Drygalski: Zum Kontinent, pp. 255–276, 336–338.

Fig. 34 Drygalski and Gazert gathering blocks of snow to produce drinking water. (Mörder, Thomas (Private possession))

Fig. 35 A provisional dog camp on the ice, photographed on March 9, 1902. (Mörder, Thomas (Private possession))

Fig. 36 The *Gauss* during its winter quartering in 1902, with its meteorological equipment, including a wind mast (*left*) and a precipitation gauge (*right*). (Mörder, Thomas (Private possession))

(who had little to do, because of the lack of ice-free rocks in the surroundings), and by the three officers of the ship as observers during their vigil.

In addition, Hans Ruser developed a tide gauge, using the ship's resources (Fig. 37). It was attached to the jib boom of the *Gauss* and, for the first time, allowed tidal observations to be made far from the coast, on a sea almost 400 meters deep.

Ernst Vanhöffen regularly examined the sea life through a crack in the sea ice, using fish baskets and nets, and discovered an astonishing richness of the tiniest organisms in this cold sea (Fig. 38).[141]

Nearly 400 meters away from the ship, Bidlingmaier erected his magnetic observatories made of ice blocks and slush. To make the work easier, he was able to apprentice a technically skilled, blind passenger, Lenhart Reuterskjöld, as his assistant, who later was able to replace him completely even when it came to making difficult magnetic measurements.[142]

In mid-March, Hans Ruser fell sick and had to be treated for syphilis.[143] Gazert prescribed the captain a mercurial inunction treatment, which, according to the

---

[141] Bidlingmaier: Zum ewigen Eise, pp. 62–64.
[142] Drygalski: Zum Kontinent, pp. 47, 408.
[143] Drygalski 1901–1903a: March 13, 1902.

Fig. 37 An improvised tide gauge on the jib boom of the *Gauss*. (Leibniz-Institut für Länderkunde, Leipzig, Drygalski estate)

Fig. 38 Fishing at the geodetic gap. A fish basket is just being pulled up. (Mörder, Thomas (Private possession))

knowledge of the time, could cause general malaise but was not associated with any further disadvantages to the patient.[144] Ruser was probably in the so-called tertiary stage of the disease, which appears, at the earliest, 3–5 years after infection. In this stage, a deficient oxygen supply to the brain produces a personality change associated with performance weakness and professional misconduct, as had now emerged in Ruser's case, and this influenced his relationship with the other officers and the crew members in an extremely negative way.[145] Because of the legal consequences it entailed, the occurrence of syphilis during Drygalski's expedition was not mentioned in Gazert's health report; otherwise, Ruser would have had to be immediately discharged from duty pursuant to Article 57.4 of the German Regulations for Seamen. As a consequence, the case was never published.

Because the geologist Emil Philippi had not found employment, since there was a lack of rocks for him to study, he soon asked to go on a sledge expedition to the coast.[146] This suggestion was very much after Drygalski's own heart and so, on March 18, 1902, Philippi embarked on a 9-day discovery journey southward, accompanied by Second Officer Richard Vahsel and the Norwegian seaman Daniel Johannsen, traveling on two sledges, each loaded with 500 pounds of equipment and drawn by seven dogs. At noon, they took a break and ate "frozen food" such as bread, sausage, sardines, or chocolate. Mostly, Johannsen walked ahead and the dogs followed his trail. This way, they progressed at an average speed of 4–5 kilometers per hour. On the way, they accurately mapped out their route and gave single icebergs characteristic names (Fig. 39).

Just before sunset, between 5:00 p.m. and 6:00 p.m., Johannsen, Vahsel, and Philippi pitched their tent mostly in the lee of an iceberg (Fig. 40). "Now lively work [develops]: one of us hands out to the dogs, who [have] tiredly lain down in the snow, their well-deserved meal, stockfish or frozen seal meat. All the while a comrade fills the cooking pot with snow, another lights the petroleum cooker and scrapes the stiff-frozen content out of a tin can. A delicious dinner is ready in about one hour's time . . . much enjoyed by the hungry sledge explorers. With it, they [drink] tea, less often cocoa, often in huge amounts because the air on the ice fields is mostly very dry and causes violent thirst."[147]

On the third day of their march, Vahsel discovered a small, dark dome on the horizon—a mountain, "the first piece of solid land"—which they reached the next

---

[144] Drygalski 1901–1903b, Auszug aus Drygalskis Südpolartagebüchern, private possession Gazert, Partenkirchen: April 21, 1903; Knitschky, W. E.: Die Seegesetzgebung des Deutschen Reiches. Nebst den Entscheidungen des Reichsoberhandelsgerichts, des Reichsgerichts und der Seeämter, Guttentag'sche Sammlung Deutscher Reichsgesetze No. 19 (1894). pp. 167–168; Gazert, Hans: Gesundheitsbericht, Veröffentlichung des Instituts für Meereskunde zu Berlin, Issue 5 (1903), pp. 46–54; Gazert, Hans: Ärztliche Erfahrungen und Studien auf der Deutschen Südpolar-Expedition, in: Drygalski, Erich von (ed.): Deutsche Südpolar-Expedition 1901–1903 im Auftrage des Reichsamtes des Innern, Vol. 7, Issue 4, Berlin 1914, pp. 297–352.

[145] Poeck, Klaus: Neurologie, Berlin (1978), pp. 268–270.

[146] Philippi, Emil: Die Schlittenreisen der Deutschen Südpolarexpedition, in: Deutsche Revue 30 (1905), pp. 103–112; Drygalski: Zum Kontinent, p. 270.

[147] Philippi, Schlittenreisen, p. 107.

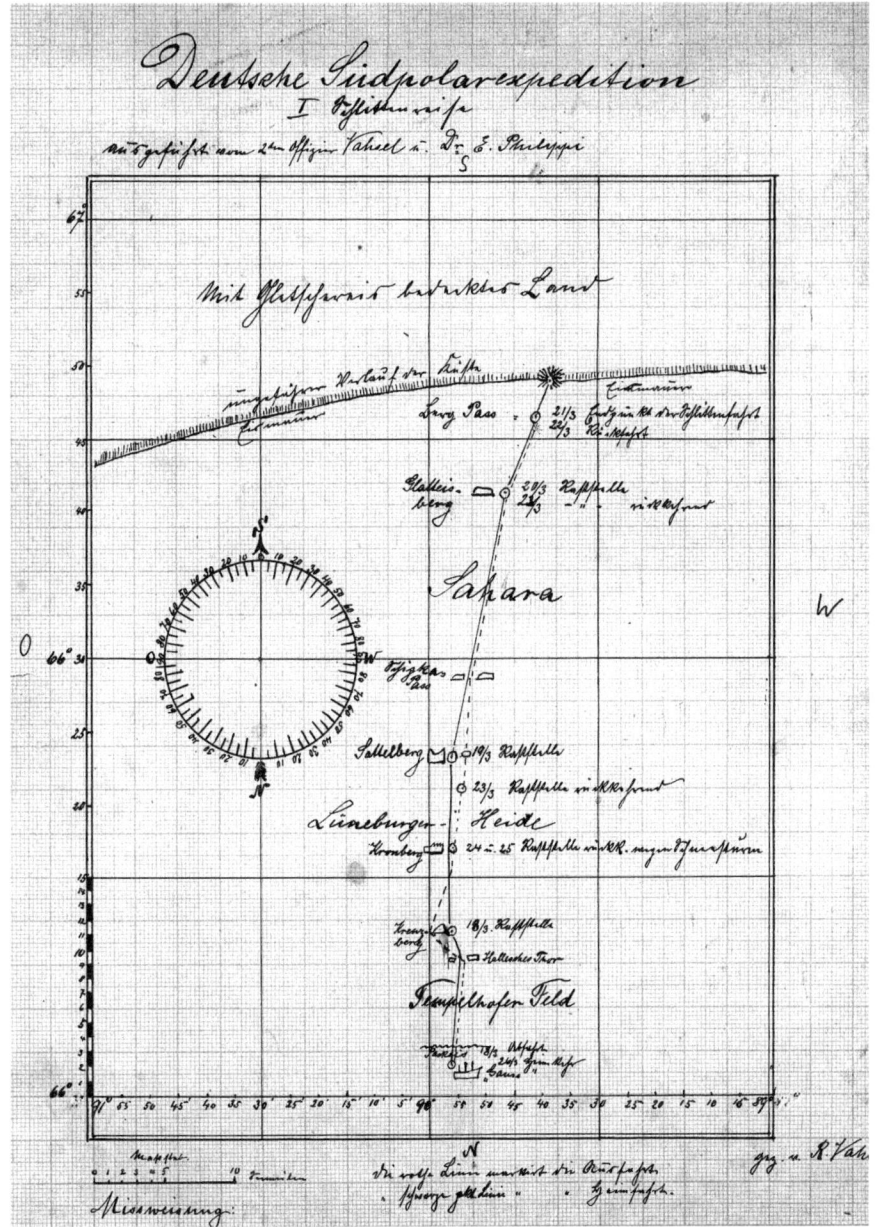

Fig. 39 The route map and name assignments for the first dog sledge journey. (Leibniz-Institut für Länderkunde, Leipzig, Drygalski estate)

Fig. 40 A view inside the cooking tent in good weather. (Mörder, Thomas (Private possession))

day (Fig. 41).[148] The 306-meter-high basalt cone, initially referred to as Black Mountain [Schwarzer Berg] and later called Gaussberg, lay directly on the border between the sea ice and inland ice. This long-extinct volcano now became the geologist Philippi's El Dorado.

While the weather in autumnal Antarctica proved to be stable enough and the first inland excursion was still in progress, preparations were made for the first manned balloon ascent close to the ship.[149] The balloon was filled with hydrogen on the Saturday before Easter. Drygalski was the first man to climb into the basket of the balloon (Figs. 42 and 43). After it had risen 50 meters, Gaussberg became visible on the horizon. At an altitude of 500 meters, the highest point of his ascent, Drygalski recognized that this was the only place free of ice in the surrounding area, which was equivalent to having found a needle in a haystack. Two further ascents with Ruser and Philippi followed, then gas had to be let out of the balloon because a sudden change in the weather was expected.

Further dog sledge travels were carried out in the coming weeks (Fig. 44).[150] On the second occasion, Philippi, together with First Officer Wilhelm Lerche and three

---

[148] Ibidem, p. 107.
[149] Drygalski: Zum Kontinent, pp. 271–276.
[150] Philippi: Schlittenreisen, pp. 108–111.

Fig. 41 Black Mountain, still a day's journey away. (Mörder, Thomas (Private possession))

Fig. 42 The winter quartering of the *Gauss* at the Antarctic Circle, from a bird's eye view, photographed in 1902. (Drygalski, Erich von (ed.): Deutsche Südpolar-Expedition 1901 – 1903: Vol. 1, table 1)

Fig. 43 A captive balloon being sent up for the first time, with great anticipation, on March 29, 1902. (Gazert, Volker (Private possession))

seamen, conducted geological investigations of Gaussberg. Their shelter, made of ice blocks, was flooded by the next spring tide.

Meanwhile, the zoological facilities were completed by leading a rope under the ship, on which a drag net was to be drawn over the ground.[151] This method was not very successful, given that the sea depth underneath the ship was 385 meters;

---

[151] Drygalski: Zum Kontinent, pp. 278–279.

Fig. 44 A tent camp with a view toward Gaussberg's northern cape. (Gazert, Volker (Private possession))

besides, the rope soon broke. However, the contraption was able to be repaired, although the ice had already begun to close in on the ship's hull.

After curious emperor penguins repeatedly appeared and could even be attracted in flocks to the ship, an emperor penguin was trained for this work. The animal was put into the water at the bow and pulled a rope fastened to its foot to a hole at the stern 50 meters away, where it buoyantly sprang out again (Fig. 45).

Drygalski led the third sledge excursion himself, accompanied by Vanhöffen, Gazert, the younger Second Officer Ludwig Ott, and three crew members, in order to survey Gaussberg with precision for a map and determine the direction and speed of the inland ice movement.[152] After they set out on April 22, 1902, the weather turned bad and at −30 °C a violent storm approached in the night, keeping the men inside their tent for two days (Fig. 46). Instead of three to four days, it took them a total of six days to reach Gaussberg. Sometime later, they had to spend four more idle days at Gaussberg because of a new snowstorm (Fig. 47).

The subsequent view from Gaussberg was tremendous and compensated for the unpleasant "surveying work on the inland ice. Under an icy wind at temperatures below −20 and −30 degrees centigrade, standing at the theodolite for hours, adjusting the telescope while the cold makes one's eyes water, handling of delicate small

---

[152] Ibidem, pp. 294–327.

Fig. 45 A penguin's unsuccessful attempt to jump out of the water. (Mörder, Thomas (Private possession))

Fig. 46 Inside the tent during the last dog sledge journey. *From left to right*: The boatswain Josef Müller repairs a sledge, Vanhöffen rests on his sleeping bag, and Gazert sits on Bidlingmaier's lap, holding the tent pole, while Drygalski is reading. (Gazert, Volker (Private possession))

Fig. 47 The tent, snowed-in on the last dog sledge journey, has to be shoveled free again after a snowstorm. (Leibniz-Institut für Länderkunde, Leipzig, Drygalski estate)

screws—everything requires both patience and endurance as hardly any other work would, as this work demands a great deal more time than it would in our climate."[153] Frostbite on the face and on fingers were inevitable occurrences. After 24 days, the sledge team returned safe and sound.

Later on, only short 1-day excursions were made on dog sledges into the nearer surroundings, which were very popular with the crew because, according to Friedrich Bidlingmaier, this added variety would invigorate both body and soul and intensify not only the relationship "between man and man, but also between man and animal."[154]

By the southern autumn (April–May), making measurements at the various observatories had become routine. When the winter came, the conditions became harder: "In the coldest months, in July and August, the thermometer was mostly between −20 and −30°C, very often between −30 and −40°C, and sometimes between −40 and −50°C. . . . However, the storms were much worse. We

---

[153] Gazert, Hans: Unser Leben im Polareis, in: Westermanns Illustrierte Deutsche Monatshefte, October 577 (1904), p. 53.
[154] Bidlingmaier: Zum ewigen Eise, p. 69.

experienced one or the other storm on sea, but such elemental forces like the one that broke loose at the edge of the Antarctic mainland are unique in their power, their duration over days and masses of snow and ice with which they completely fill out the air. . . . Soon, after leaving the ship, a pitiful crust of snow and ice formed on the cheeks, nose and eyes of the man, protecting the skin against the direct impact of the ice crystals. In his hand he desperately [held] the rope—he literally [held] his life in his hands, because he [would] hardly be able to find his way back on board without it."[155] The crew members had put up ropes between the ship and the observatories just in time for them to safely find their way to and from the ship.

After each storm, the crew had to shovel the ship free again; the next storm frustrated all previous efforts.[156] The magnetic observatories on the ice gradually began to sink into the sea because of the snow load lying on top of them.[157] When the observers finally stood knee-deep in the water, they relocated the magnetic instruments to an ice cave, which proved successful.

To bring some diversity into the dull winter period, Gazert introduced a series of lectures in which the scientists and officers reported their work.[158] To this end, all participants in the expedition, including crew members, were invited into the salon. Conviviality on board was fostered by joint grog evenings and the foundation of several clubs for singing, smoking, and playing skat.[159] In addition, there was a piano built particularly to withstand the expedition's changing climatic conditions.[160]

After the heavy winter storms had apparently abated, Drygalski set out to Gaussberg again on September 16, 1902, in company with all five scientists, as well as Vahsel and three crew members, in order to complete the measurements that had begun during the previous autumn.[161] Apparently, this was too early in time, as it took them a total of ten days to get there. The snow hut had become a ruin in the meantime, so they still had to sleep inside their tent. While Bidlingmaier and the boatswain Josef Müller dug out a cave in a snowdrift and installed a magnetic observatory for reference measurements, Vanhöffen took water samples from a tide crack in the ice and pulled up fish baskets containing all kinds of interesting creatures (Fig. 48). Stehr took care of measuring the sea ice temperatures. Gaussberg was also climbed again (Fig. 49). Three times daily, although parts of his fingers were completely stiffened by the cold, Vahsel carried out precise astronomical positioning measurements (longitude and latitude) of distinctive places at Gaussberg. The height above sea level was also determined. The distance between two sites thus defined—the so-called control points—formed the reference scale for later

---

[155] Ibidem, pp. 64–65.
[156] Gazert: Unser Leben, p. 46.
[157] Bidlingmaier: Zum ewigen Eise, p. 66.
[158] Drygalski: Zum Kontinent, pp. 345–346.
[159] Gazeri: Unser Leben, p. 46.
[160] Drygalski: Zum Kontinent, p. 655.
[161] Ibidem, pp. 400–423; Philippi: Schlittenreisen, p. 109.

Fig. 48 Noack hoisting a fish basket, using a hand winch. (Mörder, Thomas (Private possession))

Fig. 49 The entrance to the ice valley, as seen from the moraine, against the sun. (Mörder, Thomas (Private possession))

map construction. At some control points, Gazert, using a special camera connected to a theodolite, took photogrammetric pictures of Gaussberg, for which he recorded not only the locations but also the exact target directions (the compass directions and elevation angles). In addition, he targeted various distinctive sites on the mountain with a theodolite and also noted the respective target directions. Furthermore, he drew additional sketches, which later helped to identify the distinctive points in the photographs (Figs. 50 and 51). With the focal length of the camera's objective taken into account, a map of Gaussberg could then be constructed from these pieces of information, back in Germany. Drygalski himself repeated the previous autumn's trigonometric survey of the inland ice, for which he again determined the angles of the triangle grids over the ice's surface, using a theodolite.

The surveying work was finished after two weeks, and the group started on its way back.[162] On October 12, they were surprised by a storm, which confined them to their tent for two whole days, just a day's journey away from the ship. Even their warm dinner had to be cancelled because the kitchen tent had been crushed to such an extent that the cooking apparatus no longer had a sufficient air supply. Frozen bread and hard chocolate were insufficient replacements for the rice with apples they had originally planned to eat.

On another sledge journey, Phillippi, Lerche, Ott, and the seaman Karl Klück explored the inland ice in the western part of Posadowksy Bay (from October 26 to November 5), which was named the West Ice.[163]

When the crew once again complained to Drygalski about the captain on November 16, 1902, Drygalski sent Hans Ruser on a sledge to the West Ice two days later—together with Bidlingmaier, Ott, and Klück—"to cool off his temper" and make magnetic measurements and soundings there (from November 18 to November 20) (Fig. 52).[164]

On his last sledge expedition from December 1 to December 4, Drygalski wanted to get his own idea of the West Ice and, led by Philippi, was accompanied by Gazert, Ruser, and the seamen Johannsen and Reuterskjöld. As for the latter, who was one of the blind passengers, this journey was a nice reward for his previous work.[165] At the boundary of the West Ice, they sounded sea depths of more than 600 meters, for which reason Drygalski correctly assumed that the whole ice floe was floating and actually represented shelf ice.

---

[162] Drygalski: Zum Kontinent, pp. 421–422.
[163] Philippi: Schlittenreisen, pp. 109–111.
[164] Drygalski: 1901–1903a, November 16, 1902, November 18, 1902; Drygalski: Zum Kontinent, pp. 434–435.
[165] Ibidem, pp. 434–444.

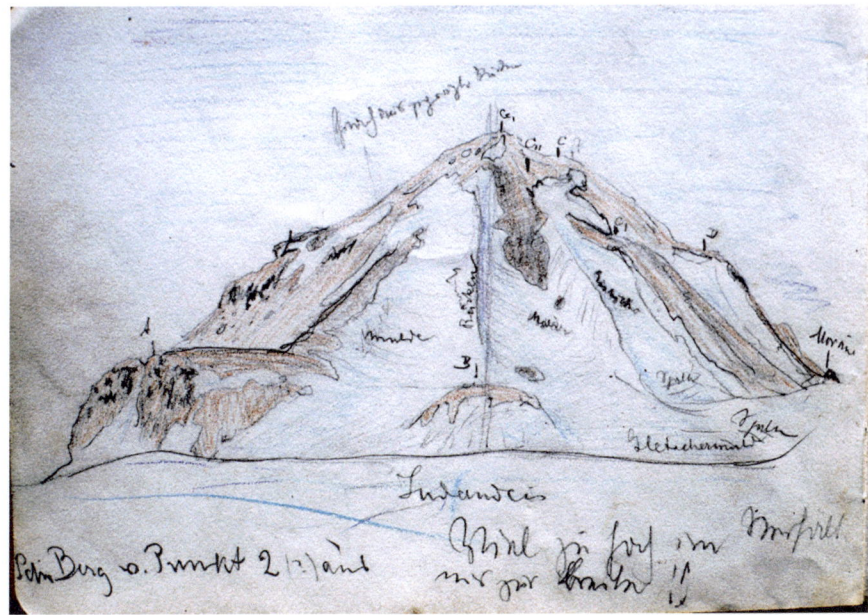

Fig. 50 Gazert's sketch of Black Mountain (Gaussberg) from Measuring Point IV, drawn on May 1, 1902. (Gazert, Volker (Private possession))

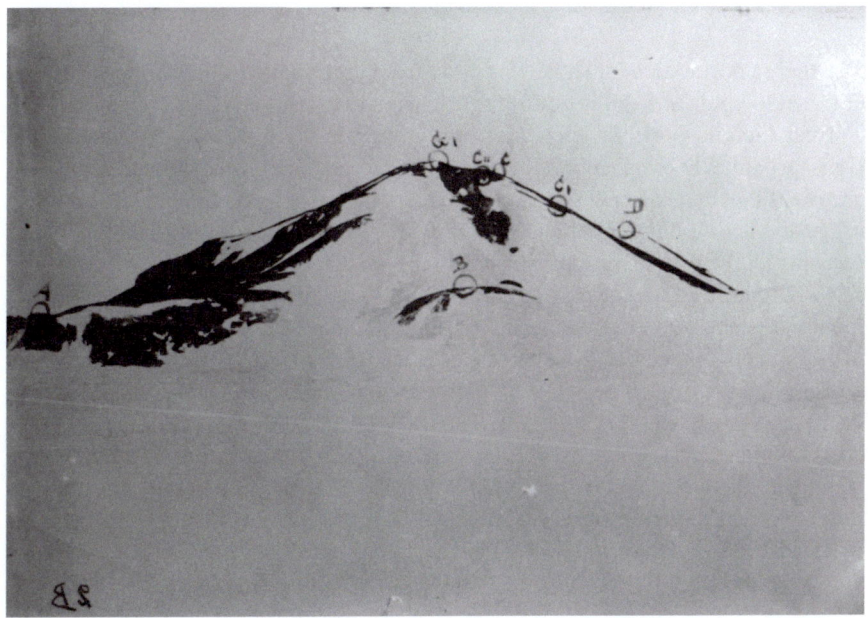

Fig. 51 Survey photography of Gaussberg. (Gazert, Volker (Private possession))

Fig. 52 Preparations for sledge travel. Sealskins in frames, hung out for drying, are recognizable *in the background*. (Mörder, Thomas (Private possession))

Other opportunities for short trips away from the crowded ship were provided by the need to hunt for fresh meat for the kitchen and to feed the dogs (Fig. 53).

Near the end of 1902, there were no indications that the ice surrounding the *Gauss* would ever break away. This fixed position was becoming an increasing concern for all on board. On the basis of his experiences in Greenland, Drygalski finally suggested laying out a 2-kilometer-long path of debris in front of the ship's bow, reaching all the way to the closest crack in the ice that was only thinly frozen.[166] The sun would heat up the dark waste and melt the ice surrounding it. In fact, within a month, this operation led to formation of the so-called Lake Titicaca, with a depth of 1–2 meters in the 5- to 6-meter-thick sea ice (Fig. 54).

Despite all of the concerns they had, the men celebrated (Fig. 55). First of all came the solstice on December 21. Three days later, on Christmas Day, presents from families and friends, as well as numerous donations, were set out before the ship under the bright southern summer sky of 1902.[167] Each crew member received three different kinds of tobacco; the members of the salon (officers and scientists) received 10 Virginias each, 10 long Dutch cigars, and 25 Havana cigars, which were

---

[166] Ibidem, pp. 443–444, 473.

[167] Ibidem, pp. 468–470.

Fig. 53 A Ross seal is dissected. (Mörder, Thomas (Private possession))

Fig. 54 "Lake Titicaca," with the latrine visible *in the background*. (Leibniz-Institut für Länderkunde, Leipzig, Drygalski estate)

Fig. 55 Midsummer celebrations in the saloon. (Geographisches Institut der Universität, Munich)

traded using a skat currency of 1/80 per point and therefore were highly welcome. On holidays, prize-shooting competitions, ice stick shooting, or kayak rides on "Lake Titicaca" were organized (Figs. 56, 57, 58 and 59).

Early in January 1903, the *Gauss* was still trapped in the ice. For this reason, the men tried to cut a path to the already ice-free fairway that had been opened by the tides, using dynamite and poles. However, these attempts remained futile.[168] Drygalski contemplated the next steps. If a message from the *Gauss* was not received in Germany by June 1, 1903, a rescue expedition would be launched to Knox Land, lying nearly 800 kilometers eastward, as had been previously agreed.[169] On the basis of the good experience gained with the dog sledges, a voyage from the *Gauss* to Knox Land would take no more than 2 months. Therefore, in March 1903, Drygalski planned to travel by dog sledge to explore the route to Knox Land.

---

[168] Ibidem, pp. 473–476.
[169] Ibidem, pp. 484–485.

Fig. 56 The geologist Philippi with a young emperor penguin. (Mörder, Thomas (Private possession))

Fig. 57 *From left to right*: Ott, Gazert, Lerche, Drygalski, and Ruser adjust their sunglasses before going hunting. (Mörder, Thomas (Private possession))

Fig. 58 Pastime pleasures on "Lake Titicaca" in single and double kayaks. (Mörder, Thomas (Private possession))

Fig. 59 Seamen on the deck of the *Gauss* on a Sunday. (Gazert, Volker (Private possession))

Fig. 60 Men working to free the *Gauss*. (Mörder, Thomas (Private possession))

In late January, the men made another attempt to remove the ice from the eastern side of the *Gauss*, using icepicks and a saw they had built on board (Fig. 60).[170] As a result of a groundswell, the ship, which had been trapped for 50 weeks, finally came free again on February 8, 1903 (Fig. 61). "We left a tremendous amount of traces behind us. Dead seals we had not retrieved, penguins, penguin heads, piles of crude fat next to them, stakes, broken dog kennels, boards, poles and cans. Where the *Gauss* had lain I still perceived upon our departure in the western snowdrift a series of dirt layers stacked on top of each other, each interspersed with tin cans, bottles, asbestos board, straw husks and other refuse, an obvious sign showing that 32 men had lived here for one year."[171]

However, because of a rising snowstorm, the *Gauss* was already at risk of being trapped in the ice again the following day.[172] The days were passing by, and the ice opened and closed again. In mid-March, the *Gauss* finally came free and could set a westerly course, where a deep advance to the south was expected to succeed because of the assumed warm ocean current. Instead, the *Gauss* would become trapped in the pack ice again two more times.

At home, no messages regarding the fate of the *Gauss* had yet been received. As early as June 1902, the Ministry of the Interior—supported by the Ministry of

---

[170] Ibidem, pp. 491–492.

[171] Ibidem, p. 506.

[172] Ibidem, p. 508.

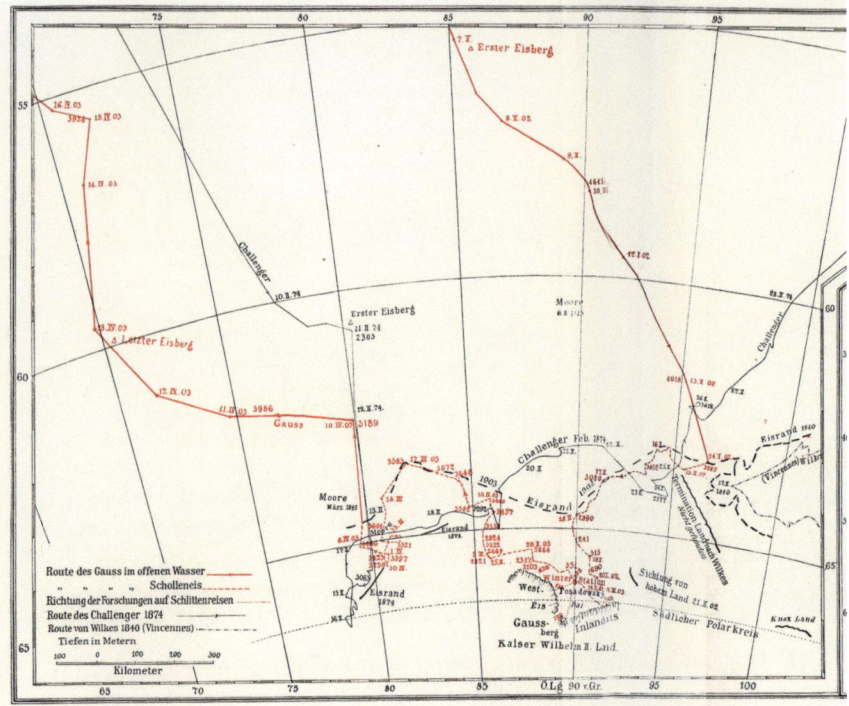

Fig. 61 The travel routes of the *Gauss* and the *Challenger* along the Antarctic coast. (Drygalski, Erich von: Zum Kontinent des eisigen Südens, Berlin 1904, pp. 254–255)

Culture, the Imperial Treasury Office, and other advisors—had already started planning a rescue expedition as a matter of routine. As had been proposed by Drygalski, Josef Enzensperger (the meteorologist at the Kerguelen Station) was to be the leader; however, he had died of beriberi early in 1903[173]—but no one at home had received notice of that yet.

In late March 1903, the first officer of the *Tanglin*, C. Neuhaus, was appointed as the captain of the rescue expedition.[174] Among the other participants who applied at the Ministry of the Interior was the physician and geographer Albert Tafel,[175] who

---

[173] Ibidem, pp. 573–575; Reichsministerium des Innern: June 11, 1902, Brief an das Auswärtige Amt, BArch Potsdam, 15.01 RMdI , No. 16154, Bl. 101-106; see also the course of the disease in Enzensperger, Josef J.: Reisebriefe und Kerguelen-Tagebuch, in: Akademischer Alpenverein München (ed.): Josef Enzensperger. Ein Bergsteigerleben (posthumous), Munich, 1905, pp. 258–275.

[174] Norddeutscher Lloyd: March 26, 1903, Brief an das Reichsministerium des Innern, BArch Potsdam, 15. 01 R MdI, No. 16155.

[175] Tafel: April 15, 1903, Brief an das Reichsministerium des Innern, BArch Potsdam, 15.01 RMdI, No. 16156.

Fig. 62 An iceberg—an unpleasant neighbor—photographed on April 1, 1903. (Mörder, Thomas (Private possession))

had attended Drygalski's first lecture on "The Geography of Polar Regions."[176] The Ministry of the Interior requested that Tafel should be ready for a rescue expedition by June 1, 1903.[177]

In the meantime, matters had escalated dramatically in the south. On April 8, the *Gauss* was caught in extreme ice pressure during a storm: "the floes rotated, the ropes we used to anchor tore, and massive ice boulders shattered before our very eyes; everything around us swayed in a wild commotion. . . . The situation was threatening, the crash could happen any moment. Hence steam was raised and the engine went on, the spanker sail was set and under the influence of both the *Gauss* wound herself through the easterly wind in a fortunate maneuver. What had often failed earlier, now succeeded in the moment of distress. . . . But our fate was clear. . . . So I gave the order to take a northerly course. It was a hard decision, surely the hardest I had ever taken, but it had to be. A firm winter [quartering] did not exist here, and new attempts to reach one were in vain at this time of the year."[178] Drygalski had realized that another advance to the South Pole would not be successful at this point and that he would have to turn back (Fig. 62).

---

[176] Drygalski: Autobiographie, p. 72.

[177] Reichsministerium des Innern: May 6, 1903, Reichsministerium des Innern: May 6, 1903, Brief an Tafel, BArch Potsdam, 15.01 RMdI, No. 16155.

[178] Drygalski: Zum Kontinent, pp. 540–541.

Back to Cape Town

Drygalski had no other choice but to reach the next telegraph office in South Africa in time to prevent the dispatch of the already prepared rescue expedition. Besides, there was now growing unrest among the crew members, who, after becoming accustomed to the calm life on the ice, now had to adjust themselves very abruptly to work on stormy seas again. Captain Ruser displayed quite abusive behavior, whereas Drygalski was in a "state of collapse."[179] Seamen and officers grumbled out loud. These massive controversies induced Drygalski to give up his initial plan to "make a short stopover in Cape Town and then spend the rest of the winter at one of the archipelagos in the southern Indian Ocean, and travel south again from there in the first days of spring."[180] Drygalski also anticipated correctly that the expedition would probably be ordered to return home after a fortunate overwintering. Yet he intended to file a comprehensively substantiated application to extend the expedition with a summer campaign and thus place the responsibility for returning south in the hands of the government.[181]

On the way back to Cape Town in mid-April, the expedition was plagued by strong westerly storms, which went along with rough seas (Fig. 63). Finally, on April 26, 1903, the volcano island of St. Paul came into view, where, on their shore leave, everyone was delighted to see something green again, albeit only moss and thorny rushes.[182] On the island of New Amsterdam, only 60 nautical miles away, they carried out further magnetic measurements and hunted wild cattle in order to get a little variety in their menu. Then they sailed, without stopping, past the Kerguelen Islands and carried out additional shipboard magnetic measurements in the central Indian Ocean. The participants in the expedition had started writing their reports even before they reached the next coast. On May 12, 1903, they encountered the Norwegian bark *Garcia*—the first ship they had seen since their overwintering—which they visited with great curiosity.

On May 31, 1903, the *Gauss* reached the South African town of Durban.[183] According to the pilot, the black plague had broken out there. For this reason, Drygalski gave the pilot his telegram addressed to the German Ministry of the Interior. The message of the successful winter quartering was received in Berlin the next day. On June 9, the ship reached Simon's Town, where, finally, everyone could go on land. There, a telegram from the Ministry of the Interior was waiting for Drygalski:

---

[179] Drygalski: 1901–1903a, April 10, 1903; Drygalski: 1901–1903b, April 10, 1903, April 12, 1903.

[180] Drygalski: Zum Kontinent, pp. 546–547.

[181] Drygalski: May 30, 1903, Brief das Reichsministerium des Innern, BArch Potsdam, 15. 01. RMdI, No. 16148; see also Dokument in Lüdecke: Deutsche Polarforschung, Anhang VI/3.

[182] Drygalski: Zum Kontinent, pp. 549–556.

[183] Drygalski: June 1, 1903, Brief an das Reichsministerium des Innern, BArch Potsdam, 15. 01. RMdI, No. 16127, Bl. 87; Drygalski: Zum Kontinent, pp. 564–568.

Fig. 63 Heavy seas on the return to Cape Town. (Mörder, Thomas (Private possession))

"[I] cordially congratulate the South Polar expedition for returning. Coming home [is] commanded as soon as [the] *Gauss* [is] ready to sail, as monetary budget [is] exhausted, Graf Posadowsky."[184]

This telegram was a hard blow for Drygalski. His reports, including his substantiated request for continuation of the expedition, were still in the post on their way to Berlin. In the scope of a summer voyage, he had intended to both collect reference data from another region of Antarctica and do justice to the nautical character of the expedition, as only short distances had hitherto been sailed in the Southern Ocean.[185] Hence, the only thing that could be done was to wait for the reply from Berlin and carry out the necessary repairs on the *Gauss* in the meantime (Fig. 64).

At this point, Drygalski learned that Enzensperger had died of beriberi. To him, it was a great loss. He had appreciated Enzensperger as the "soul" of the Kerguelen Station and had considered him highly capable of leading the potential rescue expedition because of his exceptionally practical sense.[186] The cause of beriberi was still unknown at the time.[187]

---

[184] Posadowsky: June 2, 1903, Posadowsky an Drygalski, BArch Potsdam, 15.01 R MdI, No. 16127, Bl. 89; see also Drygalski: Zum Kontinent, p. 569.

[185] Drygalski: May 30, 1903; copy in Lüdecke: Deutsche Polarforschung, Anhang VI/13.

[186] Drygalski: January 25, 1902, Brief an das Reichsministerium des Innern, GStA Merseburg, Rep. 76 Vc, Sect.1, Title 11, Part VA, No. 7, Vol. I, Bl. 418–422.

[187] Drygalski: Zum Kontinent, pp. 574–575.

Fig. 64 Second Officer Richard Vahsel. (Mörder, Thomas (Private possession))

Homeward Bound

On June 23, 1903, in preparation for the return journey, the director Max Richter chaired a discussion at the Ministry of the Interior—which included representatives from the Imperial Treasury Office, the Imperial Naval Office, the Ministry of Culture, and several professors—about the South Polar Expedition's scientific assignments.[188] With reference to the available memoranda, Richter showed that the predetermined 2-year duration of the expedition had expired and that the funding was nearly exhausted. The representative of the Treasury Office underscored the "at present extraordinarily unfavorable financial situation of the German Empire."[189] Richter telegraphed Drygalski:

> "New expedition South Polar region impossible due to exhaustion of funds. Request immediate return per order from June 2. After reception [of] instruments examine Walvis Ridge and Romanche Bay [sic!}."[190]

Walvis Ridge, west of Namibia, and Romanche Deep, already sounded on the outgoing voyage, were to be measured again on the way back home, with new

---

[188] Reichsministerium des Innern: June 23, 1903, Brief an das Reichsministerium des Innern, GStA Merseburg, Rep. 76 Vc, Sect.1, Title 11, Part VA, No. 7, Vol. I, Bl. 418–422.

[189] Ibidem.

[190] Reichsministerium des Innern: July 11, 1903, Telegramm an Deutsche Südpolar Expedition, Leibniz-Institut für Länderkunde, Leipzig, Box 74, Inventory Number 4825, Serial Number. 3.

Fig. 65 The welcome in Kiel on November 25, 1903. (Gazert, Volker (Private possession))

equipment. The *Gauss* had lost almost all of its oceanographic instruments and many miles of sounding wire in the storms; now it had to wait for replacements.[191] The crew members of the *Gauss* therefore stayed in Simon's Town for a while and made trips to the countryside. They could leave South Africa as late as on August 2, 1903.[192] On the way, Drygalski noted, "St. Helena and Ascension offered beautiful and captivating sights which we enjoyed. At last came the tropical splendor of the Azores and in-between the unveiling of some mysteries from the deep of the ocean."[193]

After a passage through Kaiser Wilhelm Canal, the first German South Polar Expedition finally arrived home, safe and sound, in Kiel on November 25, 1903.[194] The harbor seemed to be deserted, as the fleet was out on a training cruise. To welcome them, only Prince Heinrich of Prussia, the emperor's brother, had appeared, as he took a great personal interest in polar research (Figs. 65 and 66).

---

[191] Philippi, Emil: Über das Problem der Schichtung und über Schichtbildung am Boden der heutigen Meere, in: Zeitschrift der Deutschen Geologischen Gesellschaft 60 (1908), pp. 351–352.

[192] Drygalski: Zum Kontinent, pp. 610–611.

[193] Ibidem, p. 666.

[194] Ibidem, pp. 658–568.

Fig. 66 Evening festivity honoring Drygalski on February 19, 1904, at the Polytechnic Society of Stettin. (Cornelia Lüdecke (Private possession))

Kaiser Wilhelm II himself, however, did not even send a congratulatory telegram. This expressed his disappointment. He believed in "the old notion that the most important task of geographical research in a foreign continent was to remove as many white spots from the map as could possibly be done."[195] From the political point of view, the first German South Polar Expedition was a miserable failure, for Scott had been able to plant the British flag in Antarctica at 82°16.5′S, whereas Drygalski had hardly got beyond the Antarctic Circle. Thus, Germany had to concede the victory of reaching the highest southern latitude to her political rival, Great Britain. The conscientious records of measured data, the collection of geological specimens, and the multitude of biological specimens meant absolutely nothing in this context. Besides, some of the material and data required decade-long processing and could therefore not produce results as fast as a new spot on the map.[196] On the other hand, the name "South Polar Expedition" (derived from Neumayer's tradition) might have been somewhat ill chosen, since it implied that the South Pole would be the expedition's destination.

From Drygalski's scientific perspective, the position of the overwintering station, relatively far from the continent, had been fully sufficient for the intended Antarctic studies; thus, there had been no reason for more expanded sledge excursions.[197] It is no wonder he was embittered that "an enterprise in which you invested years of your life and best efforts, and then returned home conscious of having done everything together with your companions to bring the aspired goal to a good conclusion, does not encounter . . . just criticism everywhere in the fatherland which one could expect as a result of such sacrifices and successes."[198]

Back home again, the evaluation of the measurements and comprehensive biological collections had to be prepared for. Posadowsky received Drygalski at the Ministry of the Interior, saying that "he had put one-hundred thousand marks into the imperial budget, for processing of the scientific results of the expedition, so that [it] could immediately begin."[199] As employees of the imperial state, the scientists were to take care of the evaluation of the measurements, to which end they were even provided with premises of their own at the ministry in Berlin.[200]

Early in 1904, Posadowsky reported to the kaiser that unless the *Gauss* was reconstructed, it would be suited only to scientific research purposes in polar

---

[195] Creutzburg, Nikolaus: Erich von Drygalski zum 65. Geburtstag, in: Geographischer Anzeiger 26 (1925), p. 4.

[196] Drygalski, Erich von (ed.): Deutsche Südpolar-Expedition 1901–1903 im Auftrage des Reichsamtes des Innern, 20 volumes, 2 atlases, Berlin 1905–1931.

[197] Drygalski, Erich von: Allgemeiner Bericht über die Arbeiten der Deutschen Südpolar-Expedition und deren Verwertung, in: Kollm, Georg (ed.): Verhandlungen des 15. Deutschen Geographen-Tages zu Danzig am 13., 14. und 15. Juni 1905, Berlin 1905, pp. 9–10.

[198] Wagner, Hermann: Besprechung »Zum Kontinent des eisige Südens«, in: Zeitschrift der Gesellschaft für Erdkunde zu Berlin (1905), p. 346.

[199] Drygalski: Autobiographie, p. 103.

[200] Ibidem, pp. 104–105; see also Thorbecke, Franz: Die deutsche Südpolar-Expedition, in: Geographische Zeitschrift 11 (1905), p. 506.

regions.²⁰¹ As a ship made of wood would cost a lot for maintenance and lose value when decommissioned for a longer period of time, it would be advisable to sell it before long. The sale would also presumably cover the budget overrun. At last, the *Gauss* was sold to the Canadian government for the price of 75,000 Canadian dollars and assigned (under her new name, *Arctic*) to Captain Joseph-Elzéar Bernier to explore the North American archipelago.²⁰² For the time being, this put an end to any further German polar research.

Despite additional financial claims, several changes in government, the death of his publisher, World War I, and the subsequent period of inflation, Drygalski took care of complete publication of the scientific results, which were published through the commission, and at the expense of the Ministry of the Interior, between 1905 and 1931. Instead of the originally planned 10 volumes—including the expedition's findings in geography, geology, meteorology, geomagnetism, oceanography, botany, and zoology, along with two atlases containing weather charts and magnetic data—20 volumes were ultimately released because the zoological collections of 4030 species (1430 being new discoveries) filled a total of 12 volumes instead of one.²⁰³ The abundance of life in the cold polar seas was a great surprise.

The publication costs, comprising the printing costs and the salaries of the scientists, amounted, at last, to nearly the same total as the expedition itself had cost—approximately 1.5 million marks.²⁰⁴

In later years, the first German South Polar Expedition became connected with the term "Universitas Antarctica" ["Antarctic University"], which suitably referred to Drygalski's comprehensive research approach of Humboldtian tradition and remained customary usage in expert circles until late in the 1950s.²⁰⁵

---

²⁰¹ Posadowsky: January 11, 1904, Brief an den Kaiser, GStA Merseburg, 2.2.1. Geheimes Zivilkabinett, No. 21373, Bl. 124.

²⁰² Posadowsky: April 14, 1904, Aktennotiz, GStA Merseburg, 2.2.1. Geheimes Zivilkabinett, No. 21373, Bl. 138–141; see also Anonymous: Polarländer, in: Geographische Zeitschrift 10 (1904), p. 231; Anonymous: Polarländer, in: Geographische Zeitschrift 15 (1909), p. 652.

²⁰³ Drygalski: Deutsche Südpolar-Expedition 1901–1903; Lüdecke/Brogiato/Hönsch: Universitas Antarctica, p. 24. Vanhöffen had already referred to the fact that his collection could serve as the touchstone for more recent investigations. In fact, the results do constitute an inestimable reference point for current biodiversity research.

²⁰⁴ Drygalski: Autobiographie, p. 87.

²⁰⁵ Fels: Drygalski, p. 143.

Meteorology and Mutiny: The Second German South Polar Expedition (1911–1912)

The Political Framework Conditions for German Polar Research on the Eve of World War I

Ten years after Drygalski's South Polar Expedition, any collaboration between Great Britain and Germany worth mentioning no longer existed. In the meantime, the German Empire had risen to become the second-strongest sea power after Great Britain. In 1910, despite additional revenues resulting from tax reform, the navy's budget proposal was cut and fewer resources were assigned to the army as well.[1] The failure of Germany's Morocco policy in the summer of 1911 also magnified the differences between Germany and its English rival.[2] As early as 1910, Germany surpassed England's industrial production[3] and this added to the rivalry between the two countries. The heightened readiness to wage war, even temporarily, caused stock market prices to plunge.[4] Public funds were scarce; as the German Budget Act demanded that the budget surplus from 1911 had to be used to pay off old debts, new defense bills could no longer be covered financially.[5] The chancellor of the German Reich, Theobald von Bethmann Hollweg, decreed that there must be "no expenditure without coverage." The fact that imperial Germany had become the third largest colonial power after Great Britain and France[6] additionally contributed to the growing importance of the German

---

[1] Witt, Peter-Christian: Reichsfinanzen und Rüstungspolitik 1898–1914, in: Schottelius, Herbert/ Deist, Wilhelm (ed.): Marine und Marinepolitik im kaiserlichen Deutschland 1871–1914, Düsseldorf 1981, 2nd edition, p. 166–167.

[2] Lange, Annemarie: Das wilhelminische Berlin. Zwischen Jahrhundertwende und Novemberrevolution, Berlin 1988, pp. 225, 244.

[3] Kuczynski, Thomas: Das Wachstum der Industrieproduktion in den kapitalistischen Hauptländern (England, USA, Frankreich, Deutschland) und seine regionale Verteilung von 1830 bis 1913. Versuch einer statistischen Rekonstruktion, in: Jahrbuch für Wirtschaftsgeschichte, Sonderband Umwälzung der deutschen Wirtschaft im 19. Jahrhundert (1989), p. 184.

[4] Lange: Das wilhelminische Berlin, pp. 249, 450, 581–585, 875.

[5] Witt: Reichsfinanzen, pp. 171–172.

[6] Westphal, Wilfried: Geschichte der deutschen Kolonien, C. Bertelsmann 1984, pp. 50, 253.

army.[7] On this occasion, Bethmann Hollweg deliberated whether the state had reason to abandon its previous fleet policy.[8] Funding of polar expeditions by the state was a question not to even be considered in these circumstances.

Wilhelm Filchner

Wilhelm Filchner was born in Munich on September 13, 1877.[9] After the early death of his father, he first attended a private boarding school and then the Bavarian Cadet Corps. Having received his university entrance certificate, he entered military college [Kriegsschule] and was later assigned to the Bavarian 1st Infantry Regiment "König".

At the age of 23 years, he succeeded in crossing the Pamir Mountains in Central Asia on a spectacular ride during a 3-month period of leave in the summer of 1900. However, this sportive feat did not suffice to qualify him for participation in Erich von Drygalski's South Polar Expedition, for which he applied after his return.[10] He therefore decided to build a scientific knowledge base of his own for future research expeditions and studied geodesy and geography at the Technical University in Munich in 1901 and 1903.[11]

When his superior granted him the opportunity for another period of leave—for 18 months this time—Filchner planned a scientific expedition to China and eastern Tibet for cartographic and geomagnetic surveys (1903–1905).[12] To supplement his education, he took on several posts in Berlin. Among others, Ferdinand von Richthofen supported him in his systematic preparations for the expedition. Filchner could work at the Cartographic Institute [Kartographische Anstalt] and was trained in performing magnetic measurements in the Geomagnetic Observatory at the University of Berlin. He also learned astronomical position finding at the Astrophysical Observatory on Telegrafenberg in Potsdam.

Filchner's wife was Ilse Obermaier, a pharmacist's daughter from Munich, whom he married in 1902.[13] She wanted to accompany her husband on his expedition across Central Asia by all means. Following Richthofen's advice, Filchner also took along Albert Tafel, Richthofen's student, to ensure that his

(continued)

---

[7] Pogge von Strandmann, Hartmut: Nationale Verbände zwischen Weltpolitik und Kontinentalpolitik, in: Schottelius/Deist: Marine und Marinepolitik im kaiserlichen Deutschland, pp. 311–317.

[8] Witt: Reichsfinanzen, p. 169.

[9] Filchner, Wilhelm: Ein Forscherleben, Wiesbaden, 1950, pp. 9–20; Kneißl, Max: Wilhelm Filchner zum Gedächtnis, in: Zeitschrift für Vermessungswesen 82 (1957) 9, pp. 314–320; Kirschmer, Gottlob: Filchner, Wilhelm, Neue Deutsche Biographie, Vol. 5 (1961), p. 145.

[10] Filchner, Wilhelm: Ein Ritt über den Pamir, Berlin,1903; Filchner, Wilhelm: Bewerbung (Ende 1900) Leibniz-Institut für Länderkunde, Leipzig, Box 98, Inventory No. 6240, serial number 2; Lüdecke, Cornelia: Die deutsche Polarforschung seit der Jahrhundertwende und der Einfluß Erich von Drygalskis, Dissertation, in: Berichte zur Polarforschung 158 (1995), p. 44.

[11] Filchner: Ein Forscherleben, p. 46.

[12] Ibidem, pp. 47–49.

[13] Kirschmer: Filchner.

Fig. 1 Officer Wilhelm Filchner, photographed around 1905. (Filchner Archive, Munich)

wife would be better protected.[14] However, on their way, Filchner and Tafel quarreled with each other so profoundly that Filchner was to regret the consequences of this altercation as long as he lived (Fig. 1). In 1909, he and his wife divorced because she no longer wanted to "share the life of an explorer."[15]

To evaluate his studies in Asia, Filchner, with Richthofen's assistance, sought leave from military service for four years to be assigned to the Trigonometry Department at the Royal Prussian Land Survey [Trigonometrische Abteilung der Königlich Preußischen Landesaufnahme] in Berlin, where topographical maps of the empire were produced.[16] He was also able to attend the Prussian War College [Preußische Kriegsakademie] as a guest student.

Afterward, he joined the Trigonometry Department at the Royal Prussian General Staff [Königlich Preußischer Großer Generalstab], whose chief was Lieutenant General Hermann von Bertrab.[17] Following this, he led a training expedition to Spitsbergen (in 1910) and then the second German South Polar Expedition to the Weddell Sea (1911–1912).

(continued)

---

[14] Kirschmer, Gottlob: Dokumentation über die Antarktisexpedition 1911/12 von Wilhelm Filchner, in: Deutsche Geodätische Kommission, Munich, Vol. E 23 (1985), p. 25.

[15] Filchner: Ein Forscherleben, p. 82.

[16] Ibidem, pp. 80–85.

[17] Ibidem, pp. 85–86.

During World War I, Filchner was deployed first to Verdun and later to Norway and Holland, with a special commission as an intelligence officer.[18] He then moved to Berlin and began preparations to continue making geomagnetic measurements in Central Asia in order to connect the European–West Asian measurement network with the Chinese and Indian measurement network in a loop nearly 4040 miles (6500 kilometers) in length. On his expedition (1925–1928), he also visited the Kumbum Monastery, where approximately 7000 monks celebrated the Butter Lamp Festival week in December. Today, his cinema motion picture *Monks, Dancers and Soldiers* visualizes a unique Lamaist cultural asset of a time long past. Only his third expedition to Tibet (1934–1938) was supported by the state. This time, he measured the geomagnetic field between the Yellow River [Hwang He] and the Indus, combining the data with the Russian measurements.

After receiving the German National Prize for Art and Science [Deutscher Nationalpreis für Kunst und Wissenschaft], Filchner used the prize money to immediately start preparing his next expedition to Nepal (1939), where he intended to finalize his geomagnetic measurements.[19] However, on his way there in 1940, Filchner had to seek hospital treatment in British India for surgery because he had an acute kidney disease. There, he was interned because of the war raging in Europe. After closure of the camp in 1946, he stayed in Poona (now known as Pune) until 1949 in order to write his autobiography. Only in 1951 did he return to Europe, and he then lived in Zurich until he died on May 7, 1957.[20]

The Plan for a Second German South Polar Expedition

After the turn of the twentieth century, the scientific collaboration between Germany and Great Britain turned into political rivalry not only on the political level but also with regard to Antarctic exploration. It was commonly known that Scott was outfitting a new expedition to be the first to reach the South Pole.[21] Wilhelm Filchner also planned an expedition to Antarctica in order to "release all noticeably growing forces to the benefit of my country and for the glory of science."[22]

---

[18] Ibidem, pp. 144–164, 350–351.

[19] Kneißl: Wilhelm Filchner, pp. 318, 320.

[20] Filchner's estate is at the Bavarian Academy of Sciences in Munich, where his diaries, correspondence, photographs, books, and remembrances from his expeditions are kept. Hornik, Helmut/ Lüdecke, Cornelia: Wilhelm Filchner and Antarctica, in: Lüdecke, Cornelia (ed.): Steps of Foundation of Institutionalized Antarctic Research. Proceedings of the 1st SCAR Workshop on the History of Antarctic Research, Munich, 2–3 June 2005, in: Reports on Polar and Marine Research 560 (2007), pp. 52–63.

[21] Filchner: Ein Forscherleben, p. 107.

[22] Ibidem, p. 94.

He first addressed his superior, Bertrab, who got the chief of the general staff, Graf Helmuth von Moltke, interested in the expedition.[23] Since polar research was not actually a military task, Filchner received only conceptual support from this side, but it helped the preparations to gain momentum. He had to virtually start from scratch, firstly because a German polar research vessel no longer existed; secondly because governmental circles had no interest whatsoever in polar research; and thirdly because he could not expect any financial support from the empire, he had to provide the necessary funds all by himself.

Early in 1910, Filchner sent a first draft outlining the expedition to the Secret Civil Cabinet [Geheimes Zivilkabinett]—the kaiser's personal government office—in which he emphasized the national aspect of the expedition.[24] However, Wilhelm II had refused to hear anything about polar expeditions ever since Drygalski's return, so the relevant authorities also demonstrated their reluctance or, at most, merely took on mediating functions.[25] Against this background, Moltke summoned a meeting at the general staff building to give Filchner the opportunity to present his plan and a cost quotation to influential personages from the government, economy, science, and the press. At this meeting, there was encouragement for the required funds to be raised from a lottery. To start with, Filchner could win over a "faction" of interested and influential government officials, such as Culture Minister Friedrich Schmidt-Ott and Ministerial Director Theodor Lewald, both of whom had previously supported Drygalski. In addition, a committee was formed on this occasion to make further preparations.

At a general meeting of the Berlin Geographical Society [Gesellschaft für Erdkunde zu Berlin] in March 1910, Filchner informed the public about his plan to explore the relationship between the smaller West Antarctica and the larger East Antarctica.[26] His research question was: What separates these two regions—water or land (Fig. 2)?

Richthofen's successor at the University of Berlin, the geographer Albrecht Penck, recommended traveling south from the Weddell Sea and then pushing forward in a southerly direction from a base established closer to the pole.[27] If the expedition could deploy two ships, even an advancement from the Weddell Sea to the Ross Sea would be possible, although this would raise the cost of 1.2 million marks to 2 million. The Norwegian polar researcher Otto Nordenskjöld supported Filchner's plan. However, both William Speirs Bruce and Ernest Shackleton were also making plans to traverse Antarctica, starting out from the Weddell Sea.[28]

---

[23] Ibidem, pp. 94–95.

[24] Filchner, Wilhelm: (1910), Brief an Geheimes Civilcabinet, Geheimes Staatsarchiv Merseburg, 2.2.1. Geheimes Zivilkabinett, No. 21374, Bl. 5–7. On September 1, 1910, the American expedition commanded by Edwin Swift Balch was to depart to the south (Wichmann, Hugo: Südpolargebiete, in: Petermanns Geographische Mitteilungen 56 (1910), pp. 29, 150, 210; 57 (1911), p. 84). However, this expedition never became a reality. From 1908 to 1910, the French expedition was led by Jean-Baptiste Charcot to the western coast of the Antarctic Peninsula (Headland: Chronology, p. 250).

[25] Filchner: Ein Forscherleben, pp. 95–100.

[26] Filchner, Wilhelm: Plan einer deutschen antarktischen Expedition, in: Zeitschrift der Gesellschaft für Erdkunde zu Berlin 45 (1910) 3, pp. 153–158.

[27] Filchner: Plan, p. 153.

[28] Bruce, William Speirs: A new Scottish expedition to the South Polar regions, in: The Scottish Geographical Magazine 1908, pp. 200–202; Shackleton, Ernest Henry: The Imperial Trans-Antarctic Expedition, London 1914; see also Headland: Chronology, p. 263.

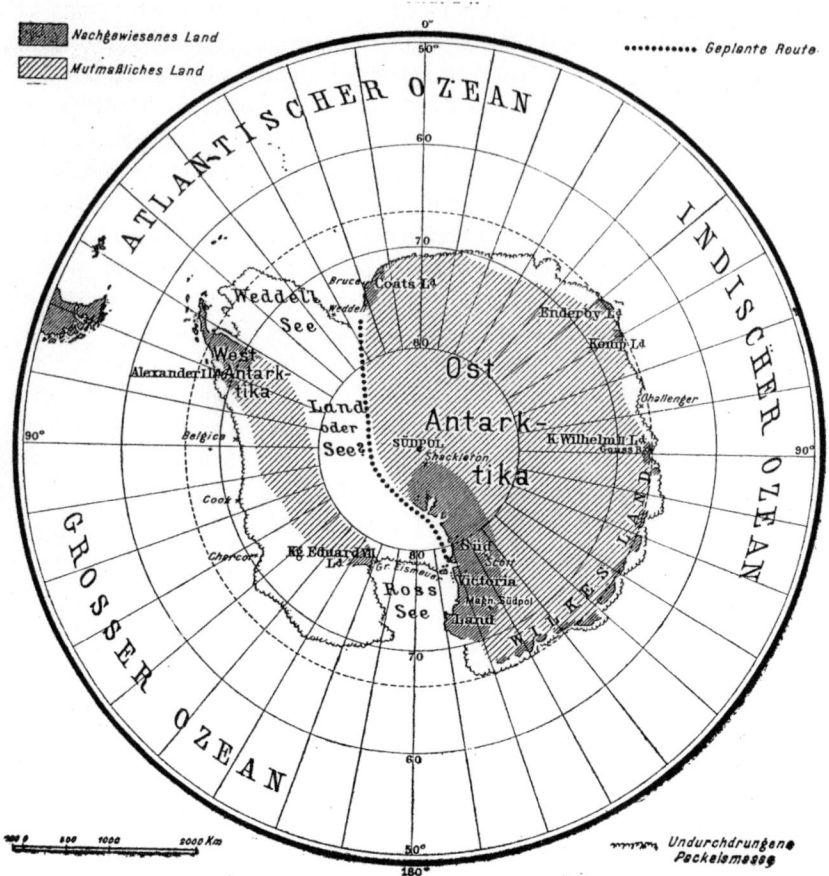

Fig. 2 A drawing of the German Antarctic expedition planned by Wilhelm Filchner in 1910. (Filchner, Wilhelm: Plan einer deutschen antarktischen Expedition, in: Zeitschrift der Gesellschaft für Erdkunde zu Berlin 45 (1910) 3, p. 154)

Provided he got the funds he needed for the crossing, Filchner intended to embark on the expedition in May 1911, allowing him, in the southern summer of 1911–1912, to lay supply depots for the southward advancement.[29] After spending the winter of 1912 in winter quarters, the traverse team was to set out, explore the region west of the pole on their way, and advance to the Ross Sea, where they would be picked up by a second ship in January or February 1913. Although Filchner did not pursue the goal of reaching the South Pole, there was much irritation about this act of "unfair competition"—of Filchner now becoming Scott's challenger. To do away with any disagreement, Filchner traveled to England in late May 1910 and met with Bruce and Scott prior to their departure.[30] They settled the problems of overlapping activities and geographical issues face to face. Although Filchner had to relinquish the

---

[29] Ule, Willi: Quer durch Süd-Amerika, Lübeck, 1924, p. 17; Anonymous: Filchners geplante Südpolarexpedition, in: Globus 97 (1910) 15, pp. 229–231.
[30] Anonymous: Notiz ohne Titel. Globus 98 (1910) 8, p. 131; Filchner: Ein Forscherleben, p. 108.

traverse for financial reasons, he hoped for Scott's collaboration and a possible encounter with him at the South Pole. With Bruce, he agreed on a geographical boundary at 20°W; the Scotsman's work area was to be the part lying to the east of it.

Filchner's plan became known in wider circles via the press. Prince Heinrich of Prussia, the brother of the emperor, summoned Filchner to his office and promised to give him his assistance.[31] As a consequence, Filchner traveled with great optimism to Norway in order to purchase the polar vessel *Björn* for the expedition. The captain of the *Björn*, Carl Julius Evensen, was hired directly after the purchase because of his previous naval experience in icy seas and because no German captain had yet shown any interest in the Antarctic expedition. In addition, Filchner also recruited Drygalski's ice pilot Paul Björvik. The *Björn* was renamed *Deutschland* and rebuilt for navigation in ice (Fig. 3).

Fig. 3 The *Deutschland* in a Norwegian dry dock, photographed in the winter of 1910–1911. (Filchner Archive, Munich)

---

[31] Filchner, Wilhelm: Zum sechsten Erdteil. Die zweite deutsche Südpolar-Expedition, Berlin, 1922, pp. 4–5, 25–26; Filchner: Ein Forscherleben, pp. 96–100.

The *Deutschland*

Technical Details of the Ship[32]

Three-masted bark with an auxiliary engine for voyages in polar seas, built in 1905
Length of the solid upper deck: 44.15 meters
Maximum breadth: 9.02 meters
Depth of hold: 5.7 meters
Engine: two-cylinder steam engine with 300 horsepower
Travel speed under sail in favorable winds: 9–10 knots (16.6–18.5 kilometers per hour)
Travel speed under steam in calm sea: 7.2 nautical miles per hour (13.3 kilometers per hour)

Special equipment on the ship

Two Lukas sounding machines
Electrical generator
Vaporization system with a connected distillation apparatus for drinking water production
Steam-powered winch
Wireless message system

A floor plan of the *Deutschland*, showing the structures on deck, can be seen in Fig. 4.

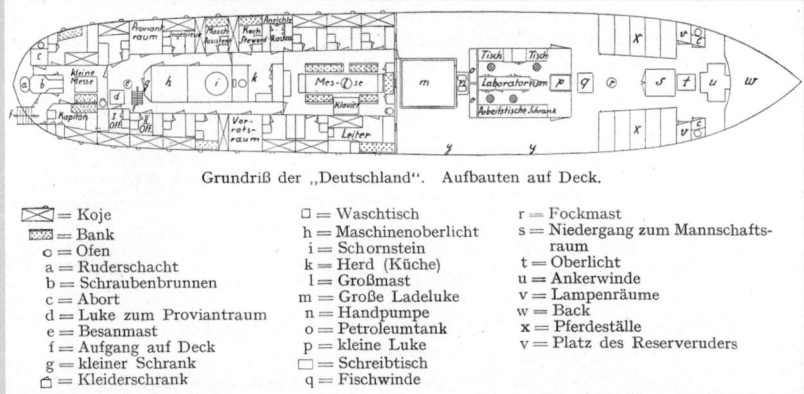

Grundriß der „Deutschland". Aufbauten auf Deck.

◻ = Koje
▨ = Bank
o = Ofen
a = Ruderschacht
b = Schraubenbrunnen
c = Abort
d = Luke zum Proviantraum
e = Besanmast
f = Aufgang auf Deck
g = kleiner Schrank
⌂ = Kleiderschrank
◻ = Waschtisch
h = Maschinenoberlicht
i = Schornstein
k = Herd (Küche)
l = Großmast
m = Große Ladeluke
n = Handpumpe
o = Petroleumtank
p = kleine Luke
◻ = Schreibtisch
q = Fischwinde
r = Fockmast
s = Niedergang zum Mannschaftsraum
t = Oberlicht
u = Ankerwinde
v = Lampenräume
w = Back
x = Pferdeställe
y = Platz des Reserveruders

Fig. 4 The floor plan of the *Deutschland*: ◻ bunk, ◻ bench, *o* oven, *a* rudder post, *b* propeller well/aperture, *c* lavatory, *d* porthole to the provisions room, *e* mizzen mast, *f* staircase to the deck, *g* small cabinet/locker, ◻ closet, ◻ washstand, *h* engine skylight, *i* chimney, *k* oven (kitchen), *l* main mast, *m* large loading hatch, *n* manual pump, *o* petroleum tank, *p* small loading hatch, ◻ desk, *q* fish winch, *r* foremast, *s* staircase down to the crew compartment, *t* skylight, *u* anchor winch, *v* lamp rooms, *w* back, *x* horse stables, *y* space for the spare rudder/helm. (Filchner, Wilhelm: Zum sechsten Erdteil. Berlin 1922, p. 32)

(continued)

---

[32] Filchner: Zum sechsten Erdteil, pp. 25–35.

The ship's crew, at the time of their departure from Buenos Aires, consisted of 21 men (Fig. 5).

Fig. 5 The crew of the *Deutschland* in Grytviken (South Georgia), photographed in 1911. *From left to right, back row*: Wolff (seaman), Olsen (seaman), Krause (ship's boy), Zäncker (ordinary seaman, later seaman), and Olaisen (seaman); *middle row*: Schwabe (first boatswain), Björvik (seaman), Dreyer (shipwright), Schalitz (second boatswain), and Engemann (stoker); *front row*: Schulze (stoker), Selle (stoker), Johnsen (sailmaker), and Hoffmann (seaman). Others missing from the picture are Klück (cook), Besenbrock (chief steward), Wilken (steward), Müller (engine assistant), Simon (second engineer), Noack (seaman and taxidermist), and Böttcher (seaman). (See also Teilnehmerliste in Kirschmer: Dokumentation, p. 4.). (Filchner, Wilhelm: Zum sechsten Erdteil. Berlin 1922, p. 27)

Training Expedition to Spitsbergen (1910)

Since the progress of the reconstruction work on the *Deutschland* was slow, Filchner used the waiting time for a small training expedition to Spitsbergen, firstly to accustom the explorers who had virtually no experience in polar regions to the rough environment, and secondly to test the measuring instruments and equipment.[33] On the advice of the Swedish geologist and Spitsbergen expert Gerard de Geer, the training ground of the hitherto unexplored area between Temple Bay [Isfjorden] and Mohn Bay in the northwest of Storfjorden was considered. The scientific objective was to survey the topography of this region (Fig. 6).

The timing was favorable because an excursion to Spitsbergen on the *Äolus* had already been planned in the scope of the International Geological Conference,

---

[33] Filchner, Wilhelm/Seelheim, Heinrich: Quer durch Spitzbergen. Eine deutsche Übungsexpedition im Zentralgebiet östlich des Eisfjords, Berlin 1911, pp. VI–VII, 39; Philipp, Hans (ed.): Ergebnisse der W. Filchnerschen Vorexpedition nach Spitzbergen 1910, in: Petermanns Geographische Mitteilungen, Ergänzungsheft 179 (1914), pp. III–IV; Filchner: Ein Forscherleben, pp. 100–107.

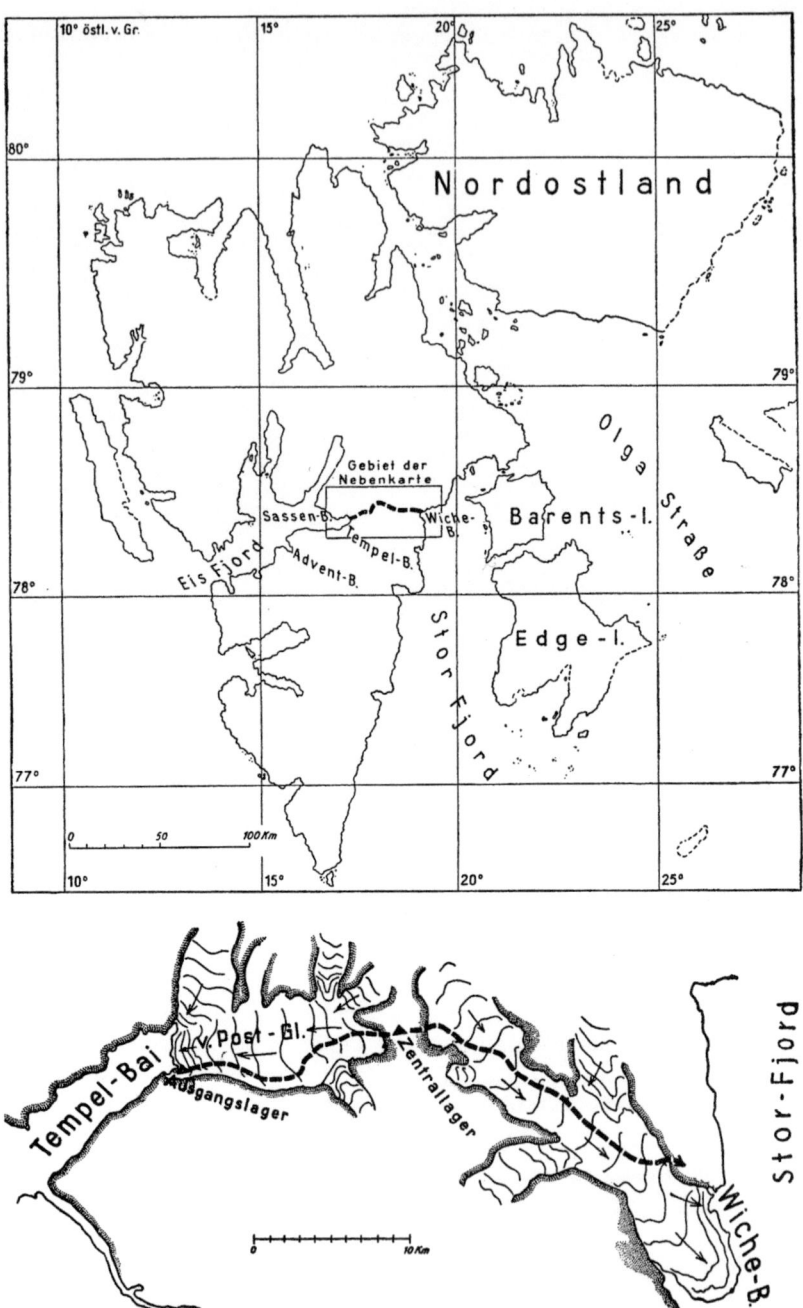

Fig. 6 A geographical overview of Spitsbergen, including the route of Filchner's training expedition. (Filchner, Wilhelm: Ein Forscherleben, Wiesbaden 1950, p. 97)

organized by de Geer in Stockholm in 1910. It was to bring part of the expedition to Spitsbergen.

The geographer Heinrich Seelheim traveled to Norway in advance to acquire two Nansen sledges (light, robust sledges developed by Nansen particularly for his traverse of Greenland) and fur sleeping bags.[34] He also bought two Lapland ponies, which were intended to pull heavy sledges because Filchner also planned to use ponies to traverse Antarctica after Shackleton's encouraging experiences. In Tromsø, everything was loaded onto the tourist steamer *Blücher*, on which the geologist Hans Philipp had come from Germany.[35] Unfortunately, the ice conditions in the summer of 1910 were quite unfavorable; the *Blücher* had already encountered drift ice at 75°40′N and failed to reach Spitsbergen.

The only thing that could be done was to return to Tromsø and meet with Filchner and his friend, the astronomer Erich Przybyllok, the meteorologist Erich Barkow, and the Graz alpinist and physician Karl Potpeschnigg before their departure on the *Äolus*. The *Äolus* was capable of taking all six participants and their equipment on board, but the ponies had to be left behind because of lack of space (Fig. 7).[36]

Fig. 7 Filchner and his companions returning from the Spitsbergen expedition in the summer of 1910. *From left to right, front row*: Przybyllok, Filchner, and Potpeschnigg; *back row*: Barkow, Seelheim, and Philipp. (Filchner, Wilhelm / Seelheim, Heinrich: Quer durch Spitzbergen. Berlin 1911, Tf 11)

---

[34] Filchner/Seelheim: Quer durch Spitzbergen, pp. 18–20, 52.

[35] Ibidem, pp. 20–41.

[36] Salomon, Wilhelm: Die Spitzbergenfahrt des Internationalen Geologischen Kongresses, in: Geologische Rundschau 1 (1910) 6, pp. 304–305; Filchner/Seelheim: Quer durch Spitzbergen, pp. 34–41.

As the luggage was not dimensioned to save weight, all of the equipment had to be adjusted for transport on sledges.[37] On August 4, 1910, the expedition landed at the Von Post Glacier [von Postbreen] northeast of Temple Bay.

The men had to harness themselves to the sledges, heavily laden with material and rations, and hoist them over partially rugged glacier faces, an effort that took a lot of strength and time. They had new experiences day after day. Exploring the route of ascent, building and dismantling the camp, melting snow to obtain water, cooking, and repairing the equipment when necessary were all quite time-consuming activities, for which reason studies other than surveying the topography were hardly manageable.

After violent storms, they finally reached the highest point of the glacier on August 11, where they built their base camp on the water divide. Carrying light field packs, one part of the group climbed down the Prince Luitpold Glacier (now known as Tunabreen), while Przybyllok and Barkow stayed in the camp to ensure accurate local positioning.

On August 18, they all returned to Advent Bay. After the shipwreck of their rowboat in the shore billows and a strenuous march with the remaining heavy gear, they finally reached the American coal mine Longyear City and traveled back to Norway on a coal steamship.[38]

The experience they gained in the far north helped Filchner and his comrades to prepare for the Antarctic expedition.[39] Only Potpeschnigg would not take part, for family reasons.

Preparations for the Main Expedition

At the 82nd Meeting of the German Friends of Nature and Physicians [Versammlung Deutscher Naturfreunde und Ärzte] in Königsberg (now known as Kaliningrad (Russia)), Filchner gave a lecture on September 23, 1910, about the Antarctic expedition he had been planning.[40] The main topic focused on the structure of Antarctica. Three theories were to be tested:

1. Shackleton and Bruce held the opinion that Antarctica was one undivided continent.
2. Nansen believed that Antarctica consisted of a group of islands and thus would be similar to an atoll.
3. Penck and Nordenskjöld assumed that Antarctica would be divided into an eastern and western part by an ice-covered arm of the sea, reaching from the Weddell Sea to the Ross Sea.

---

[37] Filchner/Seelheim: Quer durch Spitzbergen, pp. 42–113; Potpeschnigg, Karl: Verlauf und Ausrüstung der Expedition, in: Philipp, Hans (ed.): Ergebnisse der W. Filchnerschen Vorexpedition nach Spitzbergen 1910, in: Petermanns Geographische Mitteilungen, Ergänzungsheft 179 (1914), pp. 1–13.

[38] Filchner/Seelheim: Quer durch Spitzbergen, pp. 102–133, 146.

[39] Potpeschnigg: Verlauf und Ausrüstung der Expedition, p. 13.

[40] Filchner, Wilhelm: Die Deutsche Antarktische Expedition, in: Zeitschrift der Gesellschaft für Erdkunde zu Berlin 45 (1910) 7, pp. 423–430.

Preparations for the Main Expedition 87

To encourage government funding of the expedition, Graf Hugo von und zu Lerchenfeld-Koefering, a representative of the Kingdom of Bavaria, invited the kaiser and his entourage, a number of professors, and a "bloc" of governmental officials to the Bavarian embassy in Berlin.[41] Filchner reported that 791,000 marks had already been acquired and that a further 200,000 marks would be pending if a committee claimed them.[42] The kaiser listened to these descriptions with a display of goodwill; however, he had no intention of supporting the enterprise, as he did not believe in the realization of the great plan.[43]

On the same day, Filchner traveled to Munich, and in the evening, he presented his expedition plan to Luitpold, the prince regent of Bavaria. Since Filchner had formerly been invited on several occasions to the prince regent's round-table meetings at Nymphenburg Palace, he now had reason to hope for the prince regent's benevolence.[44] Indeed, the prince regent was pleased to assume the honorary patronage for his old acquaintance.[45] Shortly thereafter, a lottery to benefit the expedition was approved, which made Filchner believe his plan was financially secure and he could start purchasing the equipment he needed.

With Penck's assistance, Filchner began fine-tuning his expedition plan. In early January 1911, he composed an appeal for donations, which was circulated along with a memorandum on the "German Antarctic Expedition" (Fig. 8).[46] Filchner carefully chose the expedition's name "in order to emphasize right from the beginning that this was not an endeavor to reach the South Pole." (Fig. 9)[47] The plan was now reduced to an expedition to the Weddell Sea and, as such, was designed in accordance with the scientific standards of Drygalski's expedition.

The memorandum contained a cost estimate that now foresaw use of only one research vessel and amounted to 1.4 million marks.[48] It also included costs of 12,000 marks (mentioned under the heading "Miscellaneous") for the training expedition to Spitsbergen.

Next to Filchner and the astronomer Przybyllok, who was responsible for position finding and geomagnetics, the meteorologist Barkow, the geographer Seelheim, the

---

[41] Filchner: Ein Forscherleben, pp. 96–98.

[42] Wagner, Hermann/Drygalski, Erich von: October 29, 1910, Gedächtnisprotokoll der Sitzung von October 29, 1910, Geheimes Staatsarchiv Merseburg, Rep. 92 Schmidt-Ott, Abt. B, No. 29, Vol. 2, Bl. 167–172.

[43] In a report on Filchner's expedition, written by August von Trott zu Solz for the kaiser, there is a margin note of the kaiser referring to Solz's statement: "Filchner's expedition failed to realize the great plans [that] its leader had in mind. ... I [told] him so beforehand." Trott zu Solz, on February 7, 1913, from Trott zu Solz to the kaiser. Geheimes Staatsarchiv Merseburg, Geheimes Zivilkabinett, 2.2.1, No. 21374, Bl. 104–112.

[44] Filchner: Ein Forscherleben, pp. 44–45.

[45] Ibidem, p. 98.

[46] Aufruf: 1911, Aufruf—Eine neue deutsche Südpolarexpedition, Filchnerarchiv, Bayerische Akademie der Wissenschaften, Munich.

[47] Filchner: Zum sechsten Erdteil, p. 6.

[48] Denkschrift über die Deutsche Antarktische Expedition, Berlin 1911.

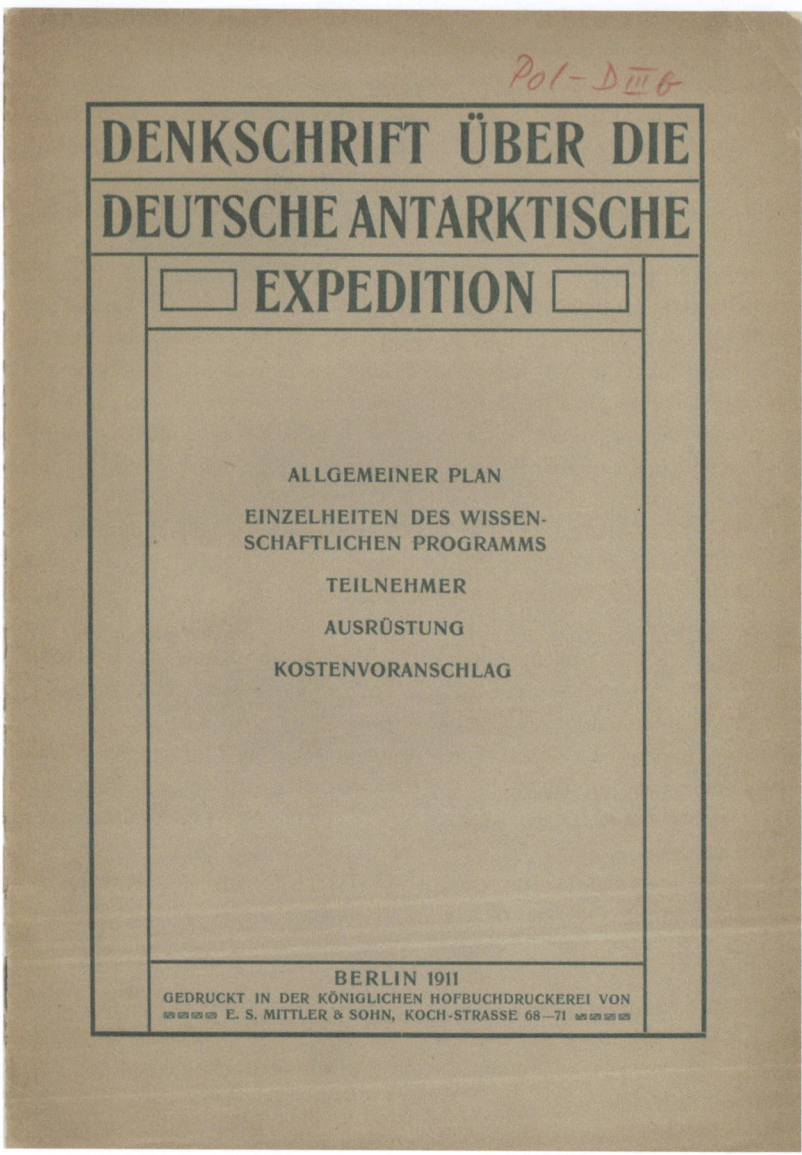

Fig. 8 The title page of the 11-page memorandum on the German Antarctic Expedition of 1911. (Lüdecke, Cornelia (Private possession))

oceanographer Wilhelm Brennecke, and the geologist Fritz Heim were employed as scientists. Kaspar Neuberger was hired as a technician to service the engine-driven vehicles that were to be taken along as a novel means of transportation.[49] The biologist Hans

---

[49] Anonymous: Die deutsche Südpolarexpedition, in: Illustrierte Zeitung No. 3534, March 23, 1911, p. 536, 538.

Fig. 9 Filchner's office sign in Berlin (Filchner Archive, Munich)

Lohmann wanted to accompany the expedition to Buenos Aires and explore the fauna and flora of the high seas on the way.[50] The men were to be joined by Ludwig Kohl (later Kohl-Larsen)—who was a splendid alpinist and skier—to serve as a physician in the land group, and by the mountaineer Wilhelm von Goeldel to do so on board the ship.[51] After the departure of the *Deutschland*, Goeldel continued to examine the food rations at home and was to catch up with the expedition at a later point in time. He was able to draw on the experiences of his alpine club comrade Gazert, the ship's physician for the first South Polar Expedition.[52] As he was unfamiliar with traveling on snow and ice, Filchner also enrolled in the Berlin chapter of the German and Austrian Alpine Club in 1910.[53]

> **The Role of the Academic Alpine Clubs [Akademische Alpenvereine]**
> Goeldel knew Gazert, as both were members of the Academic Alpine Club in Munich. Goeldel joined the club as a student of veterinary medicine in the winter semester of 1901–1902 and was elected as its second secretary for the summer semester of 1902. At that time, Gazert was already a senior fraternity member and "currently on the South Polar Expedition."[55] Albert Tafel, who joined the club in 1899–1900, was also listed as a senior fraternity member and assiduously delivered his tour reports. The annual report for 1901–1902 described the

(continued)

---

[50] Denkschrift: p. 8; Lohmann, Hans: Bericht über die biologischen Arbeiten auf der Fahrt nach Buenos-Aires, in: Zeitschrift der Gesellschaft für Erdkunde zu Berlin 1912, pp. 94–101.

[51] Filchner: Zum sechsten Erdteil, p. 8, 11, 14; Goeldel, Wilhelm von: Ueber Versuche, die Knochenregenerationsfähigkeit des Rippenperiostes nach Rippen-Resektion zu verhüten, Dissertation, Berlin 1911.

[52] Gazert, Hans: Proviant und Ernährung bei der Deutschen Südpolar-Expedition, in: Erich von Drygalski (ed.): Deutsche Südpolar-Expedition 1901–1903 im Auftrage des Reichsamtes des Innern, Vol. 7, Issue 4, Berlin 1908, pp. 1–73.

[53] Jahresbericht der Sektion Berlin des Deutschen und Oesterreichischen Alpenvereins für 1910, Berlin 1910, p. 109.

[54] Akademischer Alpenverein München: Jahresbericht des Akademischen Alpenvereins München, 1901/02, Munich 1903, pp. 6, 10, 20–21.

commemoration of the club member Enzensperger, who had died on the Kerguelen Islands during the South Polar Expedition. Goeldel had already enrolled for the winter semester of 1903–1904 as a *candidatus philosophiae* [philosophy candidate] in Berlin and was listed as an extraordinary member in Munich at that time.[56] The alpine idea had expanded to the students in Berlin, and Goeldel was among the founding members of the Academic Alpine Club of Berlin in 1903. There, in the summer of 1904, Gazert spoke "in the most interesting manner about his voyages while showing a great number of slides."[57] Finally, Goeldel, who had returned to Munich as a student of human medicine, got the opportunity also to become acquainted with Drygalski, when Gazert gave a lecture to the Academic Alpine Club of Munich on the topic "Reminiscences of Kerguelen."

During his stay in Munich, Goeldel retained his loyalty to the Academic Alpine Club of Berlin as an extraordinary member.[58]

In the winter semester of 1905–1906, he moved to Berlin for good to study there and changed his respective alpine club membership status once again.[59] Perhaps he was chosen to be Filchner's ship physician because of this connection. Also, Felix König, who was to accompany Filchner's Antarctic expedition as an alpinist, was not a person unknown to the Academic Alpine Club in Munich, for in the summer of 1906 he took part in an extensive mountaineering tour across the Ampezzano Dolomites and the Carnic Alps, of which a long description was printed in the annual report.[60]

The plan Filchner devised for the South Polar expedition included the establishment of a base station south of Coats Land for meteorological, geomagnetic, astronomical, geological, and biological studies, which was to be kept in operation for at least a year.[54] In addition, it was supposed to serve as a starting point for geographical explorations. The station would be staffed by 11 men: seven (Barkow, Przybyllok, Heim, a physician (who was also responsible for biology), a cook, and two seamen as auxiliaries) for overwintering and four for the advance to the south on sledges. If the ship froze in the ice before landing, a second advance would be attempted the following year.

The constituent assembly of the Committee for the German Antarctic Expedition [Komitee für die Deutsche Antarktische Expedition] took place in the general staff

---

[55] Akademischer Alpenverein München: Jahresbericht des Akademischen Alpenvereins München, 1902/03, Munich 1904, pp. 7, 22.

[56] Akademischer Alpenverein Berlin: I. Jahresbericht 1903/04, Berlin 1904, p. 5.

[57] Akademischer Alpenverein München: Jahresbericht des Akademischen Alpenvereins München, 1904/05, Munich 1906, pp. 4, 18, 24; Akademischer Alpenverein Berlin: II. Jahresbericht 1904/05, Berlin 1905, p. 14.

[58] Akademischer Alpenverein München: Jahresbericht des Akademischen Alpenvereins München, 1906/07, Munich 1908, p. 24; Akademischer Alpenverein Berlin: IV. Jahresbericht 1906/07, Berlin 1907, p. 41.

[59] Akademischer Alpenverein München: Jahresbericht des Akademischen Alpenvereins München, 1905/06, Munich 1906, pp. 19, 67–74.

[60] Denkschrift: p. 4.

building [Generalstabsgebäude] in Berlin on January 3, 1911, and was chaired by Prince Heinrich of Prussia, a member of the honorary board [Ehrenpräsidium].[61] Subsequent to Filchner's report on the state of the previous work, preparations were made for foundation of the Association of the German Antarctic Expedition [Verein Deutsche Antarktische Expedition], which consisted of 12 members and, under the presidency of General von Bertrab, was to render support in commercial matters.[62] The expedition ship and all other purchases that were made became the property of the association. Drygalski provided the expedition with numerous instruments.[63] Filchner also took the official directive of the first South Polar Expedition as a model.[64] However, unlike Drygalski's expedition, the participants in Filchner's expedition were in the employ not of the empire but of the Association of the German Antarctic Expedition. Their employee status considerably curtailed Filchner's authority as the leader of the expedition,[65] which its designated captain, Richard Vahsel, made drastically clear as early as in Berlin, when he stated that "according to the regulations for seamen, his capacity of captain would empower him to put Filchner in irons any time he deemed such action to be necessary."[66] It was already noticeable at this time that cooperation with Vahsel would become difficult.

New negotiations in the Reichstag for potential state support turned out to be negative once again.[67] However, since the expedition declared its willingness to consider the requirements of the Imperial Navy in its research, the ship was granted the right to hoist the imperial flag of the Ministry of the Interior, which would bring some advantages when entering foreign harbors. However, the *Deutschland* was thus subject to the authority and regulations of the Imperial Naval Office, which would curtail the expedition leader's freedom of action even further, as Filchner himself was not a member of the navy. This also included the stipulation that Filchner had to accept, for reasons of prestige, the Antarctica-experienced Vahsel (formerly a second officer on the *Gauss*) as the captain of the expedition, instead of the Norwegian captain he himself preferred. Drygalski had warmly recommended Vahsel.[68] Wilhelm Lorenzen, who had accompanied Vahsel on his voyage as the captain on a South Sea expedition from 1908 to 1910, became the first officer (Fig. 10).

---

[61] Behrmann, Walter: Polargebiete, in: Zeitschrift der Gesellschaft für Erdkunde zu Berlin 46 (1911), pp. 128–130.

[62] Behrmann: Polargebiete, p. 129; Filchner: Zum sechsten Erdteil, pp. 12–13, Filchner; Ein Forscherleben, p. 99.

[63] Filchner: Ein Forscherleben, p. 20.

[64] Anonymous: Die deutsche Antarktische Expedition, in: Zeitschrift der Gesellschaft für Erdkunde zu Berlin 46 (1911), p. 270.

[65] Filchner: Ein Forscherleben, footnote on pp. 36–37.

[66] Filchner, Wilhelm: Feststellungen, in: Gottlob Kirschmer: p. 33.

[67] Berliner Tageblatt, issue of March 21, 1911, Bundesarchiv Potsdam, 09.01 AA, No. 37564, Bl. 165; Filchner: Zum sechsten Erdteil, pp. 36–37.

[68] Denkschrift: p. 8, Filchner: Zum sechsten Erdteil, pp. 25, 99; Filchner: Ein Forscherleben, p. 99; Filchner: Feststellungen, p. 32; Drygalski, Erich von: Zum Kontinent des eisigen Südens. Die erste deutsche Südpolarexpedition, Reprint, Cornelia Lüdecke (ed.) Wiesbaden 2013, pp. 31–32.

Fig. 10 First Officer Wilhelm Lorenzen. (Filchner Archive, Munich)

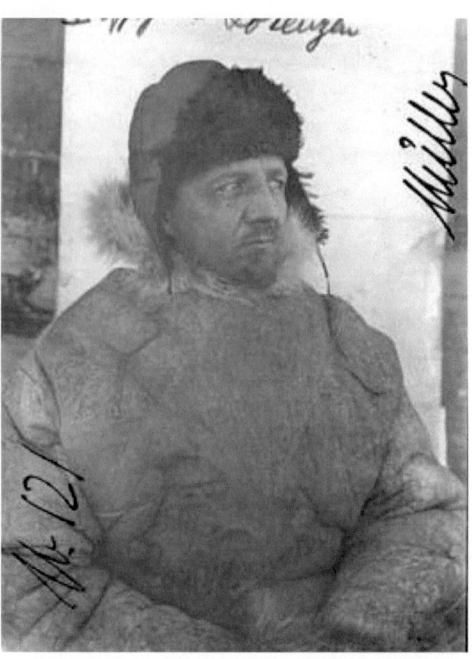

Johannes Müller was appointed as the second watch and navigation officer, and Walther Slossarczyk,[69] as the third officer, was responsible for wireless telegraph communications. In addition, Vahsel relied on human resources who had already proved their worth on the *Gauss*: the cook Carl Klück, the carpenter Willy Heinrich, and the seaman and taxidermist Georg Noack,[70] as well as the steward August Besenbrock and the ice pilot Paul Björvik.[71] A potential rescue expedition was to be led by Hans Gazert, Drygalski's expedition physician.[72]

Like Scott, Filchner ordered three motor vehicles in England to pull the heavy sledges for explorations on land.[73] The first vehicle was delivered in early February 1910 and successfully tested in the Bavarian Forest National Park (Fig. 11).[74]

The Austrian lawyer Felix König was to accompany the expedition as an alpinist. He purchased dogs in Greenland and also used the occasion to learn how to travel with dog sledges.[75] At the same time, Neuberger bargained for ponies in Siberia.[76]

---

[69] Sometimes also spelled "Slosarczyk."

[70] Filchner stated that his first name was Richard, whereas Drygalski listed him as Georg. According to the obituary, his name is Georg Richard Noack.

[71] Denkschrift: p. 8; Filchner: Zum sechsten Erdteil, p. 12.

[72] Filchner, Wilhelm: October 4, 1911, Verabredungen für die Hilfsexpedition, Bundesarchiv Potsdam, 09.01. AA, No. 37565, Bl. 100–104.

[73] Behrmann: Polargebiete, p. 130.

[74] See also Anonymous: Die deutsche Südpolarexpedition, p. 538.

[75] Filchner: Zum sechsten Erdteil, p. 8, 22.

[76] Filchner: Ein Forscherleben, p. 107.

Fig. 11 Repair of a motor vehicle in snow in the Bavarian Forest in 1911. (Filchner Archive, Munich)

The animals were taken care of at Tierpark Hagenbeck in Hamburg until the departure of the *Deutschland*.

The Departure to Antarctica

On May 3, 1911, the second German Antarctic Expedition on the *Deutschland*, under the leadership of Seelheim, left Hamburg to take on provisions in Bremerhaven.[77] There, the expedition had its formal farewell ceremony on May 6 (Figs. 12 and 13). Filchner returned to Berlin in order to make the final arrangements.[78] The first part of the voyage to Buenos Aires was dedicated to exploration of the Mid-Atlantic Threshold (now known as the Mid-Atlantic Ridge) (Fig. 14).

However, the oceanographer Brennecke had to share the time available to him for measurements with the biologist Lohmann, who conducted plankton studies and concomitantly instructed the physician Kohl. The latter was to continue biological research in Antarctica after Lohmann left the expedition in Buenos Aires.[79]

---

[77] Ule: Quer durch Süd-Amerika, pp. 17–20.
[78] Filchner: Zum sechsten Erdteil, pp. 36–46.
[79] Brennecke, Wilhelm: Ozeanographische Arbeiten der Deutschen Antarktischen Expedition. I.–III. Bericht, in: Annalen der Hydrographie und Maritimen Meteorologie 39 (1911), pp. 350–353, 464–471, 642–647; Lohmann: Bericht über die biologischen Arbeiten, pp. 95–97; Lohmann, Hans: Untersuchungen über das Pflanzen- und Tierleben der Hochsee, zugleich ein Bericht über die biologischen Arbeiten auf der Fahrt der »Deutschland« von Bremerhaven bis Buenos-Aires in der Zeit vom 7. Mai bis 7. September, in: Veröffentlichungen des Instituts für Meereskunde an der Universität Berlin, Neue Folge, A. Geographisch-naturwissenschaftliche Reihe 1 (1912), pp. 1–92.

Fig. 12 The departure from Bremerhaven. Filchner is seated *at the front left*, and Captain Vahsel is standing *in the middle*. (Filchner Archive, Munich)

Fig. 13 The *Deutschland* leaving the port of Bremerhaven on May 3, 1911. (Joester, Erich (Private possession))

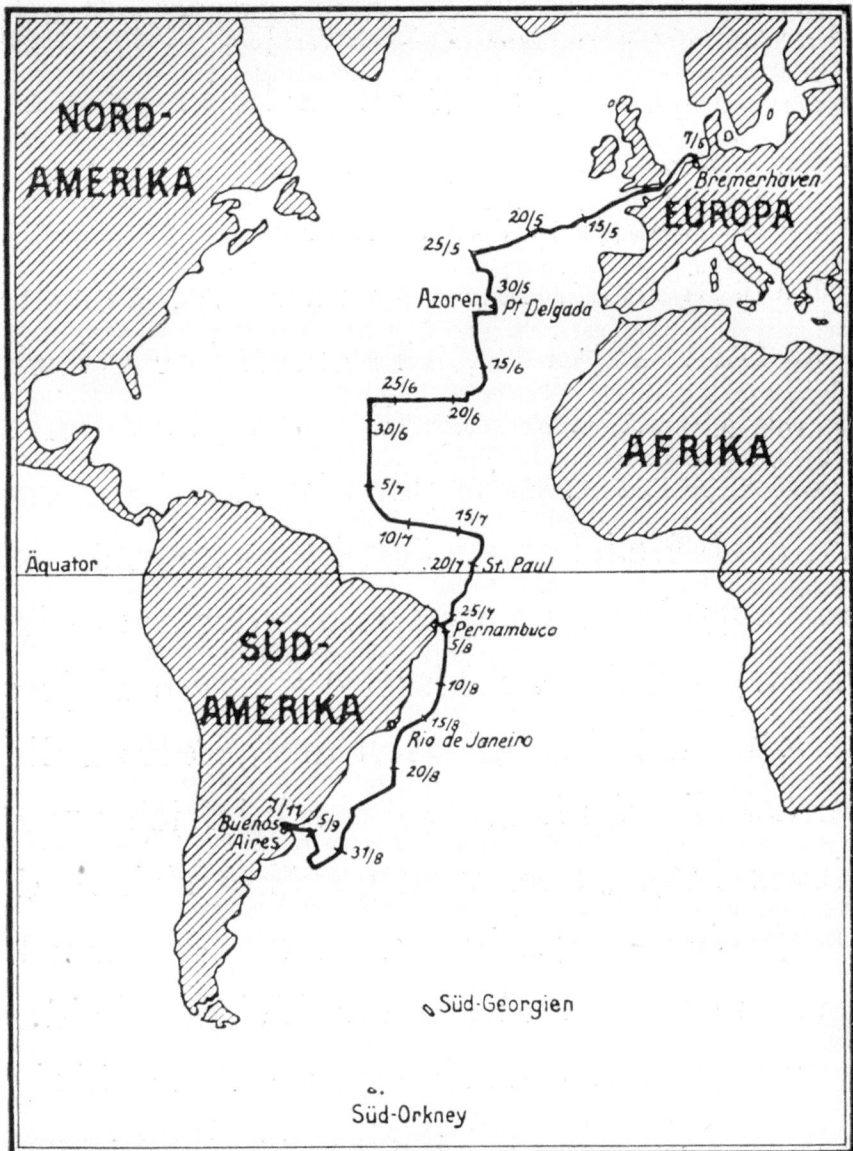

Fig. 14 The course of the *Deutschland* from Bremerhaven to Buenos Aires. (Filchner, Wilhelm: Zum sechsten Erdteil. Berlin 1922, p. 42)

A total of 175 plankton catches from various ocean depths were preserved in formaldehyde, and 226 water samples were examined under a microscope until late into the night. In addition, meteorological measurements were made regularly, for which a

Stevenson screen had been built on the stern of the ship. On the way, the geographer Willi Ule, a friend of Filchner, determined the color of the sea.[80]

On the journey, everyone soon began to criticize the tasteless food, which was most often prepared from meat conserves and referred to as "hawser" [Kabeltau] because of its texture, disintegrating into filaments.[81] Fresh fish or even fresh dolphin meat were therefore always quite welcome.

On July 6, when the ship was traveling at high speed in the open sea, the carpenter Heinrich went overboard.[82] Although the captain appeared immediately on the bridge, it seemed to take ages until they brought the ship about and were able to lower a lifeboat into the high swell. Fortunately, Heinrich was a good swimmer and managed to hold out until he could be pulled out of the water 20 minutes later. Shore leave on the rocky St. Paul Island (on July 20) also caused some excitement, for, when the crew had almost overcome the surging billows, Przybyllok fell into the water and was able to be saved only at the last moment from being attacked by approaching sharks. After rapidly collecting some rock specimens and zoological samples, everyone returned safe and sound to the homely-appearing expedition ship. The next day, they crossed the equator, an event that was naturally coupled with a line-crossing ceremony (Fig. 15).[83]

On July 26, 1911, the *Deutschland* reached Pernambuco in northeast Brazil, where Ule planned to leave the expedition and carry out other assignments in South America.[84] Heinrich Seelheim, who was actually supposed to stand in for Filchner until they reached Buenos Aires, also had to leave the expedition with a heavy heart, presumably the consequence of "expedition madness," as he described it, which "easily led to misunderstandings, disputes and quarrels among those involved."[85] The backdrop for this was that soon after their departure, there had already been differences between him and other expedition members, especially with the captain, into which the surviving records unfortunately give us no insight. Vahsel communicated to Filchner by radio that he refused to continue the voyage with Seelheim on board.[86] Filchner thereupon searched for another captain and, on his passage to Buenos Aires, recruited an officer from the *Kap Ortega*, Alfred Kling. However, after Vahsel chose to stay, Kling came on board as a navigator for the main advance on land, and from South Georgia onward he assumed the post of a watch officer.[87]

---

[80] Ule: Quer durch Süd-Amerika, pp. 16, 54–84; Ule, Willi: Bericht über geographische Studien, in: Zeitschrift der Gesellschaft für Erdkunde zu Berlin (1912), p. 102.

[81] Ule: Quer durch Süd-Amerika, pp. 30–33.

[82] Renner, Erich: Ludwig Kohl-Larsen—Der Mann, der Lucy's Ahnen fand. Lebenserinnerungen und Materialien, Landau/Pfalz 1991, pp. 70–71, Björvik, Paul: undated, Schilderung einer Reise in die Weddell-See mit der Filchnerschen Expedition 1911/13 von Paul Björvik. Translation in Archiv für deutsche Polarforschung, Alfred-Wegener-Institut, Bremerhaven, pp. 2–3.

[83] Filchner: Zum sechsten Erdteil, p. 44.

[84] Ule: Bericht über geographische Studien, pp. 102–107. Ule was the only expedition member who was also a member of the Expedition Committee. Whether he had bought himself into the expedition with a sum of money could not be ascertained. In Buenos Aires, on October 19, Ule embarked on his return voyage, as planned.

[85] Filchner: Zum sechsten Erdteil, pp. 36, 44–45.

[86] Filchner: Feststellungen, pp. 34–35.

[87] Filchner: Zum sechsten Erdteil, p. 6.

Fig. 15 The line-crossing ceremony. (Joester, Erich (Private possession))

Irrespective of Seelheim's discharge, the disagreements between the expedition members continued until the *Deutschland* reached Buenos Aires on September 7.

Filchner had already arrived in Argentina ten days earlier in order to make final preparations.[88] In particular, he made arrangements with the Argentinian government to extend the radio station on New Year's Island [Isla Año-Nuevo] in South Patagonia, which was to get a range of 1400 kilometers and be ready to operate by early 1911. Filchner also agreed with the Argentinian Weather Service [Oficina Meteorologica Argentina] that Przybyllok would be allowed to make reference measurements at the

---

[88] Ibidem, pp. 47–50.

Central Observatory in Pilar near Buenos Aires. The ponies that Neuberger purchased in Siberia and the dogs could be housed free of charge for ten weeks and three weeks, respectively, at the Zoological Garden.

Lohmann left the ship in Buenos Aires; his studies were to be continued by Kohl. Four crew members were also replaced; among them was Heinrich, Drygalski's carpenter and diver.[89] In addition, the technician Kaspar Neuberger stayed behind, allegedly for health reasons, so the purchased engine-driven vehicles were not even brought on board. The ship's physician Goeldel and the alpinist König came on board as new members.[90]

Negotiations started in Argentina obviously because of the preceding disputes among the expedition members, which culminated in duel challenges being delivered to some of the members. For example, a gentleman in a frock coat and top hat officially made a call on Kohl in his narrow cabin on board, where he "in utterly proper formality" delivered his challenge. As the physician showed no interest in dueling with pistols in a small forest, the matter was settled by a court of honor. Even though the disputes could be settled, the social cohesion among the expedition members was at great risk even before the expedition had actually begun.

To everyone's surprise, the *Fram* moored alongside the quay next to the *Deutschland* on September 12 because she had conducted oceanographic measurements in the Atlantic while Amundsen overwintered.[91] Filchner inspected the refurbished and modernized Norwegian polar research vessel. One day before her departure, 39 adult dogs and 14 puppies were taken on board the *Deutschland*[92], while Kling was to ship the ponies and the rest of the equipment directly to South Georgia on an Argentinian steamer.[93] Four powerful hurrahs from the *Fram* accompanied them as they left the port on October 4.

According to his own judgment, Filchner actually became conscious of the occurrences in Buenos Aires and other slanderous allegations only on the day of departure. He seriously considered sending the Association of the German Antarctic Expedition a telegram informing it of his resignation, as it seemed questionable that he could carry out a successful expedition with such opposition among the participants.[94] Vahsel had already lied to him on several occasions, so Filchner assumed that he would be plotting in conspiracy with Tafel, his old arch enemy. When they were on the high seas, an "awfully messy incident had taken place" in Filchner's cabin, with explicit calendar pictures depicting "various kinds of women," scarves and boots on the walls, and a stinking stockfish placed inside the gramophone funnel, all expressive of an intention to bring Filchner into disrepute.[95]

---

[89] Lohmann: Bericht; Filchner: Zum sechsten Erdteil, p. 46.
[90] Renner: Ludwig Kohl-Larsen, p. 72.
[91] Filchner: Zum sechsten Erdteil, p. 50.
[92] Björvik: Schilderung einer Reise, p. 4.
[93] Filchner: Zum sechsten Erdteil, p. 51.
[94] Filchner: Feststellungen, pp. 37–39.
[95] Barkow, Erich: Tagebuch des Meteorologen Erich Barkow an Bord der Deutschland (Deutsche Antarktische Expedition 1911/12 unter der Leitung von Wilhelm Filchner), private possession Joester, Bremen, p. 11.

Fig. 16 The ship's physician Wilhelm von Goeldel. (Filchner Archive, Munich)

The fair weather ended after three days. Southerly winds and waves arose, which additionally overlaid the existing high swell; thus, Filchner and several scientists got seriously seasick.[96] When the sea calmed down again, together they discussed the planned scientific work on South Georgia, which had belonged to the British Empire since 1908.[97] Several whaling stations were located on this subantarctic island located nearly 1400 kilometers east of Tierra del Fuego. Filchner's prime target was Royal Bay, where the geomagnetic surveys performed at the German station during the International Polar Year (1882–1883) were to be repeated for purposes of comparison, in order to determine whether there had been any alterations in the earth's magnetic field during the preceding 30 years. Current meteorological measurements were to serve for drawing comparisons with data collected at the nearby Argentinian weather station in Cumberland Bay. From there, the alpinists König and Goeldel wanted to traverse the island (Fig. 16).[98] At this point, Kohl distinguished two different tendencies among the scientists. One deemed every stop for soundings[99] or plankton studies to be unnecessary and was eager to explore the Antarctic mainland, while the other supported the oceanographic survey work. In one of Filchner's postings during those days, Kohl was announced as the leader

---

[96] Filchner: Zum sechsten Erdteil, pp. 52–53.
[97] Barkow: Tagebuch, p. 15.
[98] Renner: Ludwig Kohl-Larsen, pp. 72–73.
[99] See also Müller, Johannes: Reiseweg und Lotungen, in: Annalen der Hydrographie und maritimen Meteorologie 1912, Table 7.

Fig. 17 Depth sounding. (Joester, Erich (Private possession))

of the sledge journey toward the South Pole, which made him mighty proud. He had already begun, in collaboration with König, to have Müller explain to them the astronomical positioning methods. To this end, the height of the sun at noon was measured with a sextant, and the longitude and latitude were calculated by using the precise time.

On October 16, the *Deutschland* was near 40°S and 40°W and, in three days, was to reach the so-called Dincklage Shoal, lying 500 nautical miles northeast of South Georgia.[100] But in the early morning, Kohl collapsed with acute appendicitis, so a course had to be immediately set to South Georgia, from where he could be brought as fast as possible to Buenos Aires.[101] Przybyllok was now to go on land in South Georgia to make measurements, and König would also go along with the dogs, while the *Deutschland* would explore the Sandwich Islands until Kohl got well again and returned (Fig. 17).

The next morning, Kohl suffered another acute seizure, and so the ship's doctor decided to carry out an emergency operation.

Fortunately, there was no rough swell. The dining table in the messroom was prepared to serve as an operating table. Vahsel and Lorenzen—who, as naval officers, had knowledge of first aid—made preparations to assist, together with König and the steward Johann Wilken. After two hours, the wound was closed. Whether it is true that König's pince-nez fell into the patient's still-open wound, as Lorenzen claimed years

---

[100] Filchner: Zum sechsten Erdteil, pp. 53–54.

[101] Ibidem, pp. 54–55; Barkow: Tagebuch, pp. 16–19.

Fig. 18 Whales on the slip of the whaling station at Grytviken (South Georgia) in about October 1911. (Joester, Erich (Private possession))

later, remains to be proved.[102] Either way, the sterility that would prevail in a regular operating theater did not exist by any means. For safety reasons, guards were assigned to look after the patient.[103] The task was then to get Kohl to South Georgia as fast as possible for recovery. Filchner also decided to stay behind. According to Erich Barkow, Filchner dreaded the additional sea voyage to the Sandwich Islands.

On October 21, the west coast of South Georgia and its large mountain massifs and huge glaciers came into sight. The *Deutschland* entered the harbor at the whaling station of Grytviken belonging to the Compañia Argentina de Pesca, an Argentinian large-scale fishing operation (Fig. 18). Since their departure from Hamburg, they had traveled 12,648 nautical miles, including 1802 nautical miles since Buenos Aires.

The whaling station was run by the Norwegian Carl Anton Larsen, Nordenskjöld's captain of the Swedish Antarctic expedition on the *Antarctic*, which had sunk in the Weddell Sea on February 12, 1903.[104] The crew were able to take refuge on Paulet Island and overwintered there. Vahsel and Filchner received the latest information about the ice conditions in the Weddell Sea from Larsen and from Jörgensen, whom Filchner had not succeeded in appointing as the captain of the expedition and who now commanded the

---

[102] Renner: Ludwig Kohl-Larsen, pp. 73–75.

[103] Filchner: Zum sechsten Erdteil, pp. 55–61.

[104] Headland: Chronology, pp. 233–234, 239.

Fig. 19 The government administration building in Grytviken (South Georgia). (Joester, Erich (Private possession))

whaling vessel *Thula*.[105] Larsen, with whom Filchner soon became friends, welcomed the expedition with open arms, as it brought some change to the monotonous daily routine on this remote island.[106] For Kohl, who was brought to Larsen's house to receive further medical care, the South Polar expedition ended in South Georgia (Fig. 19).

Larsen entrusted the small steamer *Undine* and all of her crew to Filchner for a cartographic survey of South Georgia's coastline.[107] This way, by traveling on the *Undine*, Filchner was able to drop off Przybyllok and König for geomagnetic and meteorological observations at the former German station in Royal Bay (Fig. 20). In the meantime, the *Undine* sailed to Sandwich Bay (Gold Harbor) and visited Fortuna Bay on the way back.

The second voyage of the *Undine* was to explore the southern coast. On October 26, it visited King Haakon Bay, where the glaciation of the southern coast was seen to be most intensive. Here, a total of 12 separate glaciers streamed down to the sea. On the southeastern end of the island, the expedition members surveyed a nearly unknown bay, which they named Drygalski Fjord.[108] From November 1 to November 14, the

---

[105] Björvik: Schilderung einer Reise, pp. 8–9.

[106] Filchner: Zum sechsten Erdteil, pp. 59–60.

[107] Filchner, Wilhelm: Bericht des Expeditionsleiters Dr. Wilhelm Filchner, in: Zeitschrift der Gesellschaft für Erdkunde zu Berlin 47 (1912), pp. 84–89; Filchner: Zum sechsten Erdteil, pp. 73–77, 97–110; Brennecke, Wilhelm: Ozeanographische Arbeiten der Deutschen Antarktischen Expedition (Buenos Aires–Süd-Georgien–Süd-Sandwich Inseln), in: Annalen der Hydrographie und Maritimen Meteorologie 40 (1912) 3, pp. 124–131.

[108] Filchner: Zum sechsten Erdteil, p. 126.

Fig. 20 On the *Undine*. *From left to right*: Felix König (alpinist), Johannes Müller (second officer), Erich Barkow (meteorologist), Wilhelm Brennecke (oceanographer), Fritz Heim (geologist), Wilhelm von Goeldel (physician), and Wilhelm Filchner (expedition leader). (Joester, Erich (Private possession))

Deutschland went on a research cruise to the Sandwich Islands, which, in part, consisted of active volcanoes.[109] The voyage was hindered by a spell of bad weather, with a storm of up to Beaufort force 11 and waves up to 20 meters high. Even pouring oil on the waves failed to calm the sea surface close to the ship. Finally, the plan to survey the Dincklage Shoal had to be abandoned because of bad weather conditions near Zavodovski Island (Fig. 21).

On its way back, the *Deutschland* set a course to the Huisvik whaling station in Stromness Bay (South Georgia), where all were allowed to recover for a few days while 120 tons of coal was taken on board. The *Deutschland* also landed in Grytviken in order to load the timber for the winter quarters hut, which had been prefabricated in Hamburg.

On November 26, in the evening, Third Officer Slossarczyk did not come back from a fishing tour.[110] A search was initiated immediately but in vain. Only on the third day did they discover his rowboat out on the open sea, yet the man remained missing. In the meantime, Filchner discovered, in Slossarczyk's cabin, a letter to the captain, in which he had announced his suicide.[111] The reason was that Slossarczyk had got infected with syphilis in Buenos Aires, and Filchner, meaning well, had advised him to return home

---

[109] Filchner: Bericht des Expeditionsleiters, pp. 87–88; Filchner: Zum sechsten Erdteil, pp. 112–125.
[110] Filchner: Zum sechsten Erdteil, pp. 127–128.
[111] Björvik: Schilderung einer Reise, pp. 9–10. A medical report on the suicide from November 28, 1911 (signed by Goeldel, Kohl, and the station physician at Grytviken, Dr. Michelet) and a report about Slossarczyk's death on April 17, 1912, are at the Filchner Archive in Munich.

Fig. 21 The *Deutschland* searching for protection against a stormy wind of Beaufort force 7–8, 2.7 kilometers away from Zavodovski Island. (Joester, Erich (Private possession))

for treatment. Obviously, Slossarczyk had taken the disease and its consequences, which also had an impact on his future professional career, very much to heart, for the regulations for seamen demanded immediate suspension from duty in such cases.[112] This would have included loss of his polar bonus, which, amounting to one third of his annual income, would not have been insignificant.[113] The additional pay not only was intended as compensation for the special physical efforts required on a voyage to the unknown, but also was the most important incentive to embark on such a voyage in the first place. A cross raised above Trypot Bay (formerly known as Boiler Bay, and now known as King Edward Cove) in South Georgia still remembers Slossarczyk today (Fig. 22).

On December 3, the steamer *Harpune*[114] brought not only Kling (who now stood in for Slossarczyk), the 12 Manchurian ponies, and mail from home, but also Larsen's

---

[112] Knitschky, Wilhelm Ernst: Die Seegesetzgebung des Deutschen Reiches. Nebst den Entscheidungen des Reichsoberhandelsgerichts, des Reichsgerichts und der Seeämter, in: Guttentag'sche Sammlung Deutscher Reichsgesetze 19, 2. extended and improved edition, Berlin 1894, pp. 167–168.

[113] Rack, Ursula: Sozialhistorische Studie zur Polarforschung: anhand von deutschen und österreich-ungarischen Polarexpeditionen zwischen 1868–1939, Dissertation, Wien 2009, pp. 92–93, 97. The page numbers of the internet publication of the dissertation vary: http://othes.univie.ac.at/7081/1/20090730_8303884.pdf, pp. 138–139, 145. The salary tables of the expedition members ibidem.

[114] Spelled *Harpun*e or *Harpon*.

Fig. 22 The memorial cross for Slossarczyk ("Slosarczyk" is written on the cross) near Grytviken (South Georgia), photographed on February 19, 2018. (Lüdecke, Cornelia (Private possession))

daughter Margit, whom the expedition members had already got to know in Buenos Aires.[115] Erich Barkow had remarked that "how everyone was chasing after her."[116] As the *Harpune* left for Buenos Aires just before Christmas, Kohl started to leave for home. However, he still remained attracted to South Georgia. In 1913, he married Margit Larsen, changed his name to Kohl-Larsen, and returned in the southern summer of 1928–1929 to explore the interior of South Georgia.[117]

In Antarctica

While Amundsen was reaching the South Pole on December 14, 1911, the participants in the second German Antarctic Expedition saw their first ice, three days after their departure from South Georgia and nearly 4000 kilometers away from Amundsen.[118] Two days later, Barkow released the first balloon to determine the wind direction, which he followed through the telescope of the theodolite. Przybyllok read the vertical angle, Heim the time of day, and Goeldel the compass direction, while Filchner wrote down all

---

[115] Barkow: Tagebuch, p. 43; Björvik:, Schilderung einer Reise, pp. 8–9. Filchner: Zum sechsten Erdteil, pp. 129–130.

[116] Barkow: Tagebuch, p. 44.

[117] Renner: Ludwig Kohl-Larsen, p. 77; Kohl-Larsen, Ludwig: An den Toren der Antarktis, Stuttgart, 1930.

[118] Filchner: Zum sechsten Erdteil, p. 143; Headland: Chronology, p. 253.

Fig. 23 The *Deutschland*, frozen in sea ice, in March 1912. (Joester, Erich (Private possession))

of the data. Later they encountered the first table icebergs, which projected between 15 and 30 meters out of the water.[119]

Soon the *Deutschland* had to find her way through ice floes in the drift ice belt, where they encountered seals and penguins for the first time (Fig. 23).[120] The expedition members hunted the seals and used their meat to feed the dogs. Seal blubber was later also burned in the engine room to save coal.

The Christmas celebration was adapted as much as possible to the rituals of home. They all gathered around a decorated Christmas tree in the messroom, and there were presents for everyone.[121] After supper, more presents were raffled. On that evening and the following one, plenty of alcohol was consumed.

A census conducted at the year's end revealed that a total of 117 living beings were on board the *Deutschland*: 33 expedition members, one cat, eight horses, and 75 dogs (among which were 24 puppies and 13 young dogs) (Fig. 24).[122]

On January 6, 1912, Barkow flew the first box-shaped kite to which a device registering air pressure, temperature, and humidity was attached in order to investigate the meteorological conditions in the high atmospheric layers (Fig. 25).[123] Soon they crossed

---

[119] Filchner: Zum sechsten Erdteil, p. 140.

[120] Ibidem, pp. 144, 148–149.

[121] Barkow: Tagebuch, p. 57.

[122] Filchner: Zum sechsten Erdteil, p. 150.

[123] Barkow: Tagebuch, p. 61.

Fig. 24 The "menagerie" on board the *Deutschland*. (Joester, Erich (Private possession))

Fig. 25 Meteorologist Erich Barkow. (Joester, Erich (Private possession))

Fig. 26 Detail of Filchner's map showing the "Antarctica (The Sixth Continent)" and the travel route of the *Deutschland*. (Filchner, Wilhelm: Zum sechsten Erdteil. Berlin 1922, map (detail))

the polar circle and, from then on, were moving further south than Drygalski's *Gauss* had reached.[124] It was a great nautical success when the *Deutschland* sailed beyond the southernmost point of 74°25′S reached by James Weddell nearly 90 years previously on his ship, since no one before Weddell had ever managed to pass this point by ship.[125] The decrease in the sounded depths indicated that they were approaching land. After almost 1800 kilometers in drift ice, the relieving call was sounded: "Land in the southeast!"[126] At 78°S, they discovered a snow-white escarpment about 40 meters high and inland ice lying behind it, which they named Prinzregent Luitpold Land (now known as the Luitpold Coast) (Fig. 26).

They followed the ice coast further in a southerly direction, where a bay came into view that appeared to be appropriate for landing. To the west of it, a steep ice barrier continued, which swung far to the west and to which Filchner gave the name Weddell Barrier (now known as the Filchner Ice Shelf). The bay was named after Captain Vahsel and was surveyed for its suitability for winter quarters. Starting from this point, Filchner's intention was to "determine the orientation of the barrier mass towards Antarctica's

---

[124] Filchner: Zum sechsten Erdteil, pp. 153–175.

[125] Björvik: Schilderung einer Reise, p. 11.

[126] Filchner: Zum sechsten Erdteil, pp. 175–197.

Fig. 27 Break-off of inland ice with strong cornice formation in Vahsel Bay. The overhanging snow deposition on the steep leeward side blocks access to the continent. (Joester, Erich (Private possession))

inland and, if possible, to enforce the connection with Shackleton's research on Ross Barrier and the central highland region by dispatching sledging trips."[127] He thus outlined the route across the South Pole. A first landing group composed of Heim, Kling, and the seaman Olaisen examined the 8- to 10-meter-high ice wall in the southwest of the bay, while Filchner, Goeldel, and König explored the south–southwestern transition to the inland ice, where they discovered an ice-free spot of land called Bertrab Nunatak.[128] Although the captain thought Vahsel Bay [Vahsel Bucht] was a sheltered anchoring site for a ship, he saw that it was impossible to move the material needed for overwintering onto the inland ice from there. Consequently, Vahsel wanted to explore the ice barrier further westward, hoping the sea ice would break up completely. However, after they saw no alternative to climbing onto the ice, they returned to Vahsel Bay, where the *Deutschland* apparently reached her southernmost point of the expedition at 77°40′S (Fig. 27).

König's new investigations revealed that the sea ice was separated from the inland ice by frozen water channels, and Heim deemed the barrier ice at Vahsel Bay suited to landing. There, a huge ice complex seemed to have already been firmly attached to the inland ice for many years (Figs. 28 and 29).

---

[127] Ibidem, p. 188.
[128] Filchner, Wilhelm: Aus den Tagebüchern von Wilhelm Filchner, in: Gottlob Kirschmer: Dokumentation, p. 92; Filchner: Zum sechsten Erdteil, pp. 204–206.

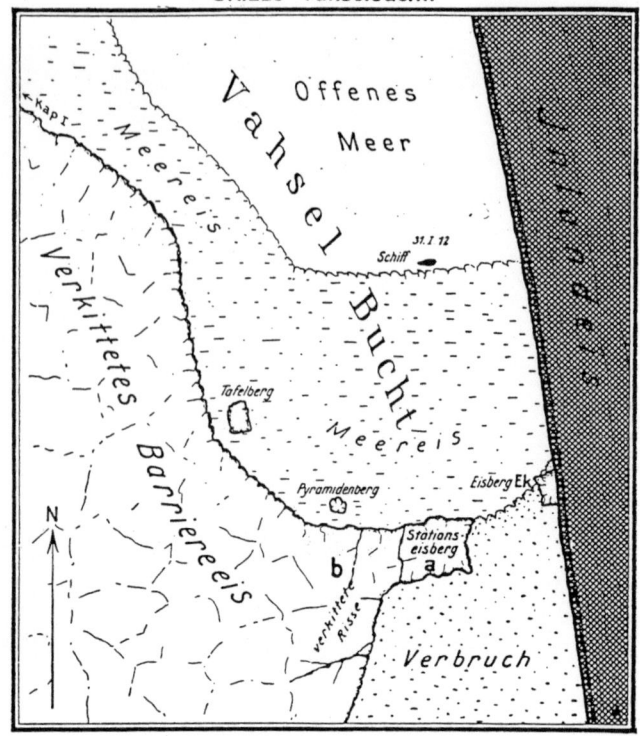

Fig. 28 Vahsel Bay and its surroundings. (Filchner, Wilhelm: Zum sechsten Erdteil. Berlin 1922, p. 190)

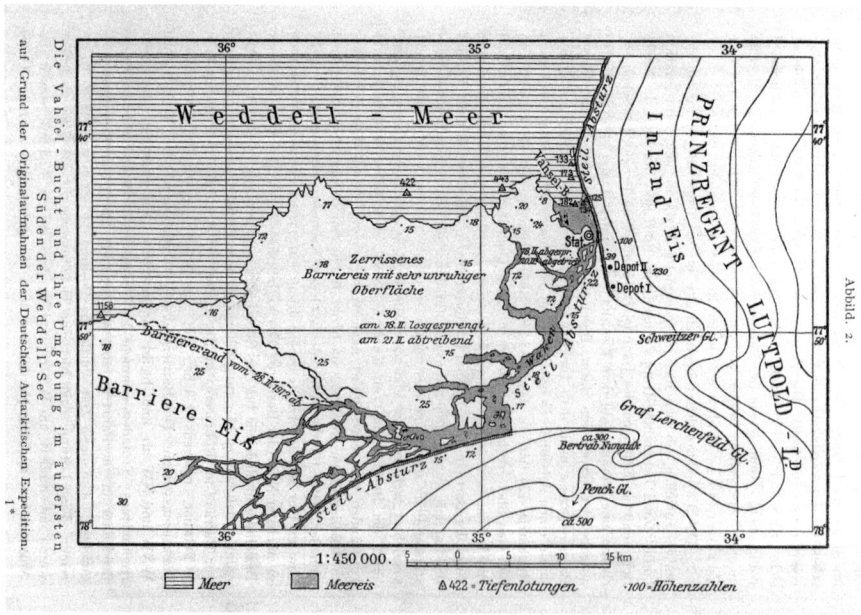

Fig. 29 The newly discovered Prinzregent Luitpold Land and the ice barrier. (Filchner, Wilhelm: Zum sechsten Erdteil. Berlin 1922, p. 211)

After Filchner had decided to move 90 tons of material close to an iceberg in the southeast of the bay to build a station on the inland ice, a suddenly rising, gusty northwesterly wind turned a 400-meter-wide strip of 2-meter-thick sea ice into a field of debris. For this reason, Filchner urged new exploration of the northwestern part of Weddell Barrier.

However, after they had failed to find an alternative, they entered Vahsel Bay again on February 5.[129] There, the narrow side of a large iceberg 12 meters high and approximately one square kilometer in area was firmly cemented to the barrier and selected to be the so-called station iceberg, as more suitable access to the inland ice could not be found. Transport to this site appeared to be quite manageable by use of a snow ramp frozen to the ice and sledges drawn by dogs and horses. The iceberg had not run aground but was floating above a sea depth of 130 meters, and was held in place only by its connection with the barrier ice. Björvik, who was employed only as a seaman but whose long experience as an ice pilot was indispensable, allegedly had not expressed any qualms about Vahsel's choice of the station iceberg; at least, this was what the captain communicated to Filchner. In his memoirs, however, Björvik stated that he had answered Vahsel's question about the location of the station with a counter-question—namely, asking why the station should not be built directly on the barrier.[130] This, however, was not possible, in Vahsel's opinion.

To be on the safe side, a food depot was first to be laid out for the construction work on the inland ice.[131] While Vahsel was in charge of the unloading and transportation of the material, Filchner oversaw the work on the winter quarters. During the night before February 7, the wind grew into a storm, pulling the ice anchors out of the sea ice and setting the *Deutschland* adrift in the bay. Two days later, the bay was free of ice and work could be continued directly on the iceberg.

Filchner had planned three sledge expeditions for the station. On the first, Filchner, Goeldel, Kling, and König were to explore Prinzregent Luitpold Land to the barrier and the potential connection with the Ross Barrier. On the second, Heim, Lorenzen, and the seaman Böttcher were to advance across the barrier ice in the south to 60°W and determine the continuation of the mountain range on Graham Land to the south or southeast. The third expedition was dedicated to exploring the inland ice in Prinzregent Luitpold Land toward the east–southeast, with Vahsel having to choose the three participants.

On February 9, after all of the required commodities had been swiftly loaded on the ramp to the iceberg, the ponies were fetched (Fig. 30). The animals took great delight in rolling around in the snow and kicking wildly after the long transport, but the dogs struggled against leaving the ship.

When Filchner asked Björvik what he thought about the choice of the station's location, he answered that he had "neither heard, read about or been present, when a polar expedition was built on a floating iceberg."[132] Filchner thus learned that Vahsel had not

---

[129] Filchner: Zum sechsten Erdteil, pp. 206–222.

[130] Björvik: Schilderung einer Reise, p. 16.

[131] Filchner: Zum sechsten Erdteil, pp. 223–230.

[132] Björvik: Schilderung einer Reise, pp. 17–18; Filchner: Aus den Tagebüchern, p. 95.

Fig. 30 Material transport on the station iceberg. (Joester, Erich (Private possession))

told the truth. He reassured Björvik that the iceberg would not move and, besides, a boat would still have to be brought up there. This idea upset Björvik, who said, "either he or I am very stupid, one of us is it by all means, how does he figure a boat can be lowered from a drifting iceberg to the water if the lowest part is at least 20 meters high and as vertical as the wall of a building?"[133]

Six men, who were dwelling in a tent at the site of construction, began building the 17.5-meter-long and 9-meter-wide station house under the supervision of the boatswain Schwabe, who was the foreman. The topping-out ceremony proceeded in a hurry on February 13, with beer, cigars, and chocolate (Fig. 31).

On February 18, a tremendous noise was heard, so loud that the building crew working on the iceberg had the impression "as if hundreds of heavy pieces of artillery had fired simultaneously."[134] Since it was on a Sunday morning and work had paused, everyone remained lying idly in the tent. Although Björvik heard a steam pipe whistle and later, when he was feeding the animals behind the edge of the iceberg, saw the tip of the main mast of the *Deutschland* showing a flag, the men on the iceberg did not pursue the matter any further. Only when the navigation officer Müller appeared did they learn that the iceberg had broken loose and was now set in motion. Two days later, if the weather had been bad, they probably could not have been rescued and it would have been a death sentence.[135] Müller brought them Filchner's order to lower 15 of the best dogs down from the iceberg in the lifeboat. Fortunately, all of the horses and almost all of the

---

[133] Björvik: Schilderung einer Reise, pp. 18–19.
[134] Filchner: Zum sechsten Erdteil, p. 245.
[135] Björvik: Schilderung einer Reise, pp. 23–25.

Fig. 31 The topping-out ceremony for the winter house on the station iceberg. (Joester, Erich (Private possession))

equipment could be salvaged. Only the house, some coal, and a few other items remained behind, among which was a dog, which had broken free and could not be caught again.

At last, Filchner posted a message about the expedition inside the station house, which might never be found:

"77.45′S

34°34′W

Arrival end of January 1912. Station house drifted away due to spring tide. Still possess timber for three small houses, otherwise everything else salvaged, including dogs and horses. All is well. Will land further south on the ice barrier.

Filchner"[136]

A tidal wave up to three meters high had changed the situation in the bay completely.[137] It was the first time (and probably is still the only time) that eyewitnesses observed the break-off of a gigantic table iceberg. As a trained topographer, Filchner was the right man in the right place to meticulously record all of the changes that occurred within the next ten days on maps. At the site of the small Vahsel Bay and the broken-off floes, a new large bay had developed, which Filchner named Herzog Ernst

---

[136] Filchner: Zum sechsten Erdteil, p. 253.

[137] Filchner: Zum sechsten Erdteil, pp. 250–256.

Bucht [Duke Ernst Bay] (now known as Vahsel Bay) after his benefactor Herzog Ernst II of Saxe-Altenburg.

The compulsory inactivity on board caused Filchner great strain.[138] The drifting away of the extensive icefields had put an end to his original plan of exploring the connection between West Antarctica and East Antarctica from a fixed station. Fortunately, they had saved one third of the timber originally meant for the station house and could use it to build a small house to live in.[139]

The last storm abated as late as February 23, allowing them to anchor in Herzog Ernst Bucht. Filchner decided to lay out a depot on the inland ice for the exploration crew. However, Vahsel opposed a new landing of station material. He wanted to wait until the *Deutschland* was firmly frozen into the bay. Therefore, on February 25, Filchner, Goeldel, and Kling placed a depot on the western slope of the inland ice at an altitude of approximately 200 meters. It consisted of one sledge, three sacks of sledging provisions (each containing three daily rations for four men), and two cookers, along with two packets of matches, well sealed against moisture.[140]

For the main depot (Depot II), Filchner planned 1000 kilograms of food (calculated for four men, to last 8 months), plus 140 pieces of stockfish as dog food and 140 kilograms of maize for the ponies, 160 liters of petroleum, two tents with installations for cooking and living, four sleeping bags, two glacier ropes, and signal flags. However, the next storm was approaching and the ship had to leave the anchoring site again.

Brennecke and Heim wanted to be dropped off on the inland ice to engage in scientific work at long last. Filchner approved, but only after the second depot was laid out. Vahsel had concerns, for there was a chance that the *Deutschland* would be kept from returning into the bay after a storm and then drift off with the ice. February 28 was a splendid day, and it seemed that the main depot could be laid out three kilometers north of Depot I. This would finally provide the basis for Filchner's plans on land. However, young ice had formed in the bay overnight, and so no boat could be lowered to fetch Heim and Brennecke. After the anchor was raised, the *Deutschland* was at the mercy of the drifting ice, against which she could hardly regain control even by using engine power. All the while, one could observe the two scientists as they carried out their measurements on land. It was not until March 3 that they could be taken on board again (Fig. 32).

Considering Vahsel's concerns and fear of the ice-trapped ship's potential drift in a westerly direction, and the perilous ice pressures inevitably to be expected then, a landing was definitely not to be considered in this season. Filchner therefore decided to leave things as they were, return to South Georgia and overwinter there, then make a new attempt to push forward in the next southern summer, hence in late December. This time, a significantly smaller station and a sledge crew of only two scientists was to be brought onto the inland ice and, above all, a drift of the *Deutschland* in the Weddell Sea

---

[138] Filchner: Ein Forscherleben, p. 124.

[139] Przybyllok, Erich: Deutsche Antarktische Expedition. Bericht über die Tätigkeit nach Verlassen von Südgeorgien, in: Zeitschrift der Gesellschaft für Erdkunde zu Berlin (1913), pp. 9–11.

[140] Filchner: Zum sechsten Erdteil, pp. 264–283.

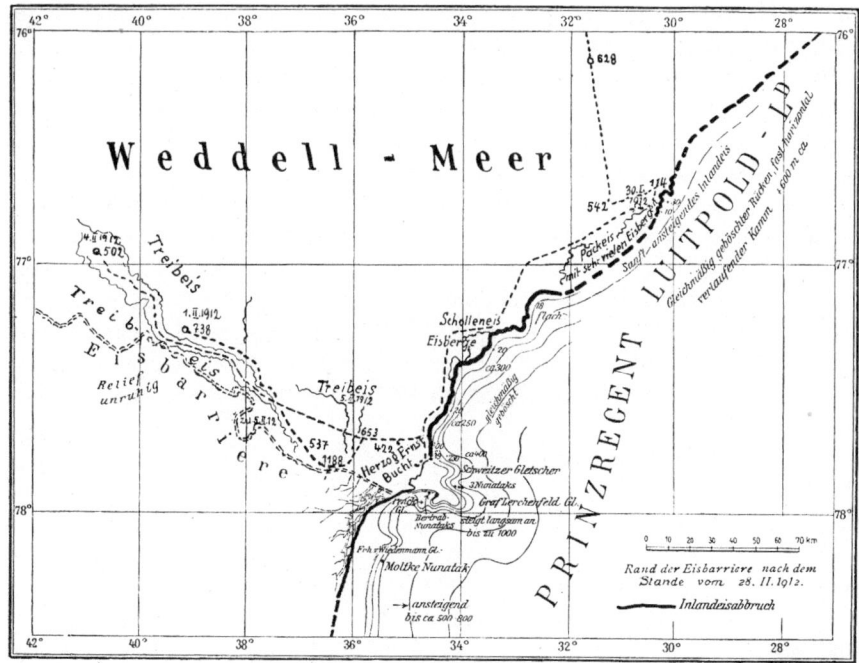

Fig. 32 The newly discovered Prinzregent Luitpold Land in the east, Herzog Ernst Bucht (newly formed after the drift of the gigantic ice floes), and the ice barrier in the west. (Filchner, Wilhelm: Zum sechsten Erdteil. Berlin 1922, p. 198)

had to be avoided. At an assembly, Vahsel explained that from a nautical point of view, the current situation would be untenable and that landing a station, no matter how small, would be absolutely impossible.[141] Still, Filchner proposed to postpone decisions on further measures until the next morning, depending on the ice conditions.

The next morning, the sea was a devastating sight to behold. "So close to the goal and back!" Filchner wrote.[142] It could not be helped; they had to leave the bay and travel northward. The young ice was gradually increasing. The situation was exacerbated by drift ice, icebergs, and ice debris, whose density grew in the course of time. A fairway was hardly visible, and they could travel only several nautical miles each day. Even occasional dynamite explosions did not help much, although the ice debris was hurled up to 20 meters into the air. But they could now melt the ice from large ice hills (hummocks) to produce drinking water, as the upper regions were as good as salt-free, thus reducing their consumption of coal to produce drinking water by evaporation, using salty sea ice.

Two days later, Barkow, Przybyllok, and Filchner discussed their work program.[143] Przybyllok wanted to set up his astronomical and geomagnetic observation huts on the

---

[141] Przybyllok: Deutsche Antarktische Expedition, p. 12.

[142] Filchner: Zum sechsten Erdteil, p. 284.

[143] Barkow: Diary, pp. 96–97.

shore, while Barkow preferred to move on to the inland ice plateau to carry out his meteorological measurements and kite ascents.

The evening's events escalated when Goeldel, acting like a madman, severely scolded the carpenter Dreyer, with whom he had celebrated First Boatswain Schwabe's birthday. Goeldel was to blame, as he was totally drunk, had started a fight with Schwabe and Dreyer, and had threatened them with a gun.[144]

The next day, Filchner called a meeting and gave them a severe reprimand. Thereupon, Brennecke flared up and refused to tolerate such generalizations, as they were not a "company of soldiers."[145] In his diary, Filchner characterized Brennecke as an uneducated and unpleasant fellow, "who had already caused much damage owing to his drinking in his cabin (before and after meetings)."[146] The tensions persisted. Filchner tried to find out who was against him and who was on his side. He also no longer trusted the captain either, with whom he had drunk a bottle of wine and spoken to on several evenings in February.

Still, they could not get out of each other's way. On the occasion of the birthday of the honorary protector of the expedition, Duke Ernst II of Saxe-Altenburg, the *Deutschland* was decorated with flags and there was a small celebration with champagne.[147] In his speech, Filchner announced that after entering the next port, "he would immediately return to Berlin and the expedition would thus come to an end. At any rate, he would make arrangements for an expedition likely to be equipped with new forces [and] new funds, which would continue our investigations. Thus the whole matter comes to an end."[148] Heim considered this to be the greatest folly the expedition leader had ever committed. Barkow even hoped that Filchner would go away and never come back. No one would notice his absence.

After they had hardly moved from their position at 73°34'S and 32°12'W, Vahsel let the fire in the huge steam boiler go out. The *Deutschland* was trapped in the ice. The small boiler, which supplied electricity for illumination, could be kept in operation by burning the blubber of one seal per day. Finally, the drift across Weddell Sea and the ship's "imprisonment in the approaching polar night"[149] began on March 15, 1912. Everyone hoped they would not have to share the fate of Nordenskjöld's *Antarctic*.[150] Since the drift in the ice went constantly westward, Filchner believed that Graham Land must be part of West Antarctica and not an island, as others on board believed. He therefore anticipated that the drift would soon switch to a northerly direction, which, in fact, did subsequently occur.[151]

---

[144] Filchner: Aus den Tagebüchern, p. 93.

[145] Barkow: Tagebuch, p. 97.

[146] Filchner: Aus den Tagebüchern, pp. 93–94.

[147] Filchner: Zum sechsten Erdteil, p. 291; Barkow: Diary, pp. 99–100.

[148] Barkow: Tagebuch, p. 99.

[149] Filchner: Ein Forscherleben, p. 126.

[150] Headland: Chronology, pp. 233–234.

[151] Filchner: Zum sechsten Erdteil, p. 314.

Fig. 33 The "aft shipmen" ["Achterschiffer"]. *From left to right*: Goeldel, Heyneck, Lorenzen, Brennecke, Heim, and Barkow. (Joester, Erich (Private possession))

Each day was just like the last from now on. The scientists informed Filchner that they requested support from the seamen, and Filchner forwarded their requests to the captain, thus avoiding certain frictions between the parties.[152] At noon, all received the same three-course meals, with four courses on Sundays and Thursdays. Until coffee in the afternoon, they were off-duty, then they resumed work until suppertime at 6:00 p.m. They hunted penguins and seals for dog food and fuel, and gradually made the *Deutschland* winterproof. In their spare time, the expedition members now made use of the abundant library. They played cards, sang songs, and made music. Filchner, however, spent most of the evenings alone in his cabin. He spoke only with Przybyllok, Kling, and König, who informed him that, astern, where Vahsel and the other officers were accommodated and plenty of alcohol was flowing in the evening, something was brewing against him, instigated by Goeldel (Fig. 33).[153] Kling divulged to Filchner that Vahsel, Lorenzen, Heim, and Brennecke had already dispatched unfavorable reports about him from Grytviken (South Georgia). Those men now stood against him, together with Goeldel—a circumstance that did not exactly improve the prospects of

---

[152] Ibidem, pp. 292–294.
[153] Filchner: Aus den Tagebüchern, pp. 94–95.

Fig. 34 The winter quarters of the *Deutschland* and scientific installations on the ice, photographed on May 9, 1912. *From left to right*: Hut for dogs, hole in the ice for measurements, absolute hut, and station hut. (Filchner, Wilhelm: Zum sechsten Erdteil. Berlin 1922, p. 321)

overwintering in harmony. Nonetheless, scientific research and ascertaining survival in the ice had to go on.

On March 26, Barkow attempted the first balloon ascent in a manual operation to investigate the meteorological conditions of the upper air.[154] It took an hour for 2300 meters of wire to be spooled off, and hauling it back in with a winch would take an equal amount of time. Later, a house was built to accommodate the inflated balloon, along with the kites and the hydrogen cylinders.[155] On April 1, 1912, the dogs were finally brought onto the ice and chained to long steel ropes anchored in the ice.[156] The ponies initially came onto the ice only in the daytime; subsequently, they would spend the night on the leeward side of the ship. Now the installation of the single measuring stations on the ice began, consisting of two so-called magnetic huts to measure the variation and the absolute values of the earth's magnetic field, and a geodetic hut (Fig. 34). In addition, the depth of the sea was sounded and oceanographic measurements taken through two holes in the ice, which were constantly kept open.[157] Explosives were stored on a hummock, and an area for winter sports was created, with a sledge run and tracks for downhill skiing on a hummock.

Brennecke was very busy during the drift, for he would constantly measure new sea depths. Only the geologist had little to do, because of the lack of rocks, but he examined ice, snow, and rime instead. With Goeldel's help, he also took many pictures, which were immediately developed in the darkroom they had at their disposal (Figs. 35 and 36).

---

[154] Barkow: Tagebuch, pp. 109–110.

[155] Filchner: Zum sechsten Erdteil, p. 307, 310.

[156] Ibidem, pp. 305–307.

[157] Ibidem, p. 303.

In Antarctica

Fig. 35 The workplace on the deck. Barkow is *second from the left* and Brennecke is *at the far right*. (Joester, Erich (Private possession))

Fig. 36 The ice structure, seen under a microscope. (Joester, Erich (Private possession))

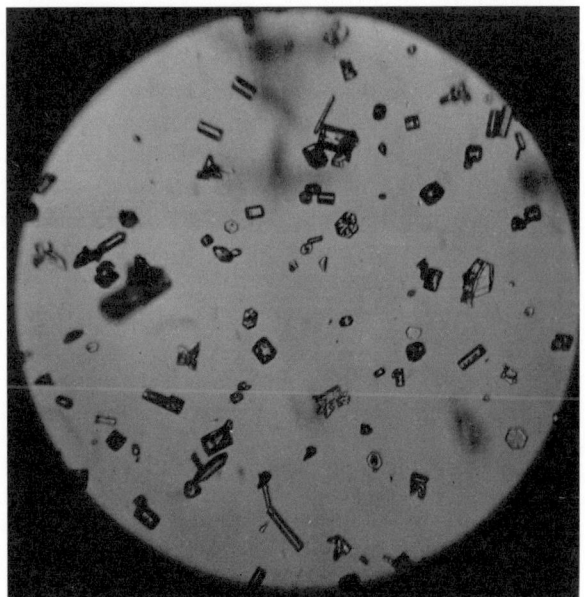

In mid-April, the first southern lights could be seen.[158] Draperies (curtain-like polar lights) rippled in bands from left to right until suddenly "a very bright ray [shot] up to the zenith." Then rays would run from right to left, vertically in relation to the horizon.

The first sledge journey took place from May 19 until May 22, led by König and including Heim and Kling, who, at the instigation of the captain and other officers, was signed off as the watch officer and was now assigned to the scientific staff.[159] The destination was a 8.5-kilometer-distant iceberg lying in a field of ice debris.[160]

In winter, the "clover leaf" (composed of Filchner, Kling, and König) took advantage of an occasion to visit Morell Land, which lay about 65 kilometers west of the *Deutschland* and had been first sighted in 1823.[161] As they set out, Björvik happened to be outside, slicing seal meat, and was the only one who bade the party farewell.[162] He had already seen plenty of starts of sledge journeys on his polar expeditions, and all had enjoyed a big farewell, especially if the leader of the expedition took part. On the 160-kilometer-long sledge journey from June 23 to June 30, it was discovered that Morell Land, charted at 70°32'S and 43°45'W, did not exist.[163]

Altogether, the feared polar night passed without any great "dreadfulness and suffering," for the weather was generally favorable, with hardly any storms. This was very fortunate, considering the tensions on board.[164] The *Deutschland* drifted "slowly, hardly noticeable, but steadily, coffined in an endless sheet of ice, melancholy, despondency spread on board."[165] In early June, the captain began to feel unwell. He had occasionally felt sick since May because of an allegedly old heart ailment (Fig. 37).[166] Filchner laconically conjectured that it was due to "the consequences of wine, women and alcohol!"[167]

The signs of the expedition's deterioration continued to grow in number. On the day of the solstice (June 21), which indicated the middle of the polar winter night, even the usual big party was cancelled.

In July, König separated fully from the *Deutschland* and built himself an igloo close by.[168] It had been too warm for him inside the ship, and it also would be a more practical solution because of his daily sledge excursions. In the course of the following months, König displayed more and more unusual behavior toward Filchner and the other scientists, who believed he had gone mad. There was not much social life in the winter. Filchner withdrew entirely, left any social event early, and hardly took part in any conversation. Inside his cabin, he planned a second advance to the south over Herzog Ernst Bucht.[169] Even on his birthday, Filchner spent the entire day in his cabin, where a few of

---

[158] Barkow: Tagebuch, p. 121.
[159] Filchner: Aus den Tagebüchern, p. 99.
[160] Filchner: Zum sechsten Erdteil, pp. 322–323.
[161] Ibidem, p. 327.
[162] Björvik: Schilderung einer Reise, p. 39.
[163] Filchner: Zum sechsten Erdteil, pp. 337–340, 349.
[164] Barkow: Tagebuch, pp. 123–1 25.
[165] Filchner: Ein Forscherleben, p. 127.
[166] Przybyllok: Deutsche Antarktische Expedition, p. 15.
[167] Filchner: Aus den Tagebüchern, pp. 101–102.
[168] Ibidem, p. 103; Barkow: Tagebuch, pp. 125–126.
[169] Filchner: Zum sechsten Erdteil, pp. 351–352.

Fig. 37 Captain Vahsel in his cabin. (Joester, Erich (Private possession))

his crew members came to congratulate him. The scientists also discussed a second journey to the south. To this end, Heim, who wanted to return home, asked Barkow to take over the geological work.[170] Brennecke also asked him to continue his oceanographic measurements, for the same reason. The meteorologist would thus be quite overburdened.

Early in August, the physician informed Filchner that Vahsel's condition had exacerbated in the course of the previous three days.[171] The captain died on August 8, 1912. The actual cause of his death was subject to professional confidentiality and was never made known publicly. Yet Goeldel told Filchner, under a pledge of secrecy, that Vahsel had confided in him about the true nature of his disease—syphilis—just a day before his passing.[172] Even before the departure of the expedition, Vahsel had told Filchner that he had never had syphilis. He would not have been allowed to take part in the Antarctic expedition if it were otherwise. Officially, "cardiac dropsy and kidney disease" was entered into the records as the captain's cause of death.[173] This spared Vahsel's relatives the financial losses they would incur if his death were officially ruled as being due to syphilis.[174] In accordance with the regulations for seamen, First Officer Wilhelm

---

[170] Barkow: Tagebuch, pp. 131–132.
[171] Filchner: Zum sechsten Erdteil, pp. 356–357.
[172] Filchner: Aus den Tagebüchern, p. 106.
[173] Filchner: Zum sechsten Erdteil, p. 357.
[174] Knitschky: Die Seegesetzgebung, pp. 162–164.

Lorenzen rose in rank to become the captain.[175] This caused tensions to flare up once more, as Filchner was fully unable to cope with the self-conscious Lorenzen. At a festive ceremony two days later, Vahsel's body was committed to the sea through a hole in the ice (Fig. 38).

In August, the previous northern drift was followed by one in a predominately eastward direction. In early September, there was strong ice pressure against the bow of the *Deutschland*. Meter-thick ice floes piled up, building an ice pressure ridge four to five meters high.

Finally, on September 7, the station ice floe broke but did not drift apart. The small boiler was then heated up again in order for the ship to be able to leave the place quickly in case there was an opening wake, i.e., an ice-free passage.

Toward the end of September, more and more cracks formed in front of the bow toward open water. At the end of the month, the ship was made ready to sail.[176]

As the chances for the *Deutschland* to be freed were getting better, the "expedition madness" began to draw ever-widening circles. During the night of October 16, Filchner learned that a murder attempt with a Browning had been made against König, who had been living in his igloo and increasingly saw himself surrounded by enemies, and that König suspected the physician Goeldel, who allegedly wanted to poison him after a sledge journey late in June.[177] Barkow believed that König was "no longer ... in his right mind" and that he "had only imagined everything."[178] At any rate, Barkow and Goeldel put the evidence to the test in front of four witnesses, confirming that all had clearly heard the two shots that were fired, whereas no one had heard any shots whatsoever during the alleged murder attempt. Apparently, everyone's nerves were on edge at that time.

As Goeldel, who was under the influence of alcohol, collided with Filchner once more in the evening, Filchner no longer slept in his bed the following nights, out of fear for his own life; instead, he spent the night on a bench with a rifle and cartridges, so that the doctor could not shoot him through the cabin wall.

In November, Filchner felt offended by Second Officer Johannes Müller because of a trivial matter, so he had Przybyllok, acting as his second, deliver him "a challenge to a duel with pistols under severe conditions."[179] Müller, however, declined and said he would rather kill himself instead. As a consequence, much to Müller's relief, Filchner invariably ate in his cabin after that.

Early in November, Barkow, apart from his pilot balloon ascents, conducted the 126th kite ascent and thus, within just one year, accomplished more than his hero, Alfred Wegener, had achieved in two years on his Danish expedition to East Greenland in 1906–1908 (Fig. 39).[180]

---

[175] Filchner: Zum sechsten Erdteil, pp. 357–359, 363, 371.
[176] Ibidem, p. 379.
[177] Filchner: Aus den Tagebüchern, pp. 112–114.
[178] Barkow: Tagebuch, pp. 172–173.
[179] Ibidem, pp. 178–181, 183–184.
[180] Ibidem, pp. 174–175.

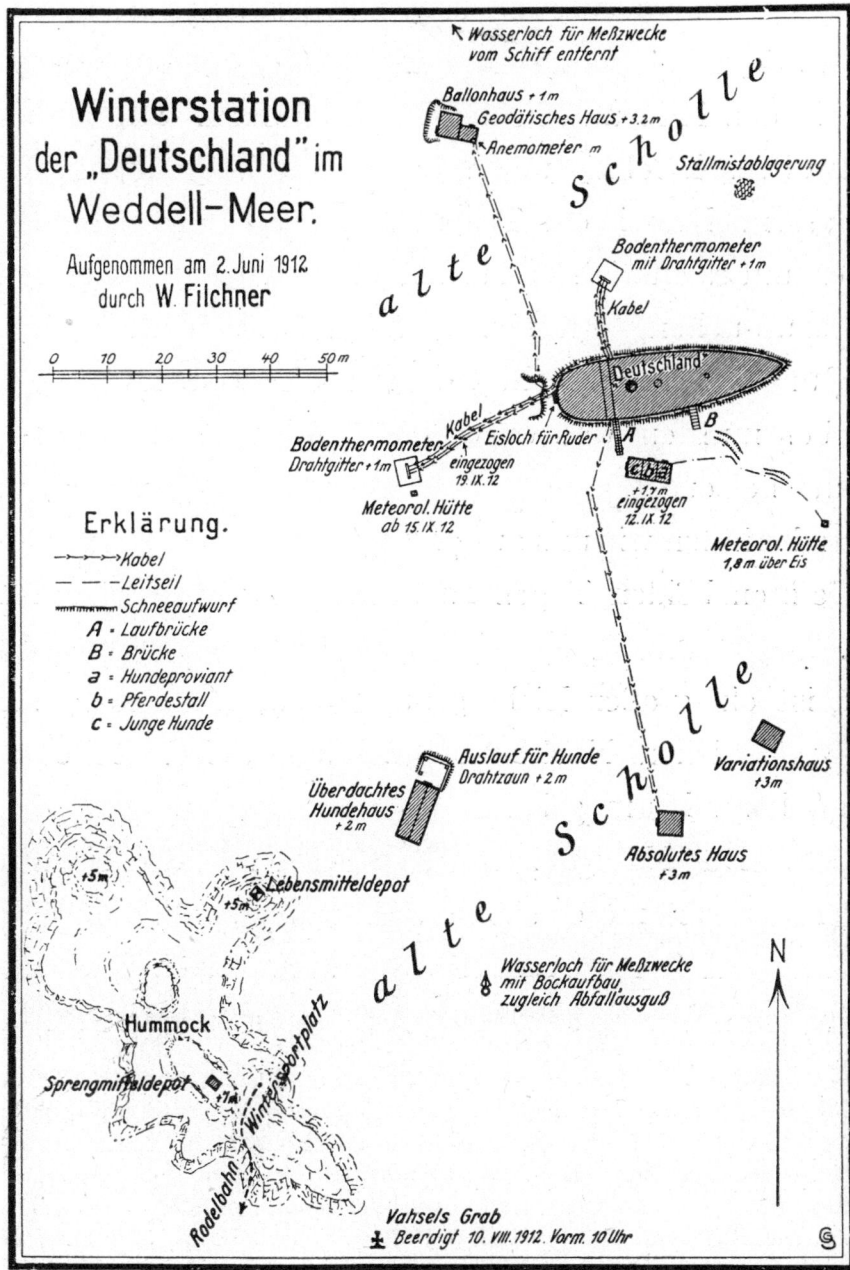

Fig. 38 A drawing showing the winter quarters of the *Deutschland* in the Weddell Sea on June 2, 1912, including all of the exterior installations and Vahsel's grave. (Filchner, Wilhelm: Zum sechsten Erdteil. Berlin 1922, p. 306)

Fig. 39 A pilot balloon ascent in Antarctica, showing the wind mast and the balloon hut *in the background*. An expedition member is getting prepared to follow the balloon with the theodolite. (Joester, Erich (Private possession))

The Return and the Scandal

On November 26, after a drift that had lasted nearly nine months, it took several explosions before the *Deutschland* could free itself from the ice at 63°37′S and 36°34′W, and set a course to South Georgia.[181] On the anniversary of the ship's departure from Grytviken on December 11, everyone received a bottle of wine from the inexhaustible stocks of alcohol.[182] Five days later, they finally left the pack ice belt behind them.[183] Barkow now experienced one of the most interesting days of the entire expedition, when, from the crow's nest on the main mast, he counted approximately 200 icebergs, each measuring about 20–40 meters in height and coming in a great variety of shapes (Fig. 40).[184] However, despite everyone's great anticipation of Grytviken, the

---

[181] Filchner: Zum sechsten Erdteil, p. 390.
[182] Barkow: Tagebuch, p. 202.
[183] Filchner: Zum sechsten Erdteil, p. 396.
[184] Barkow: Tagebuch, p. 211, 215.

The Return and the Scandal

Fig. 40 A table iceberg. (Joester, Erich (Private possession))

division in the group remained. On the evening prior to their arrival, Heim wrote a letter, in the name of Filchner's adversaries, to the station leader Larsen, stating that they by no means wished to meet with Filchner and König in the event of a likely invitation to his house (Fig. 41).

> "When the first act was over, the leader of the expedition now declared that Lorenzen would not have any command on board from now on and that he would appoint First Officer Kling to replace Lorenzen as captain, against which both the German seamen and Captain Lorenzen protested and said that the leader had no right to appoint or discharge the captain. Then it finally quieted down, so that the Englishmen could welcome the leader and speak a few words with him before they rowed back to land.
>
> Now Captain Larsen, the director of the whaling corporation in Grytviken, came on board. After he had spoken to the leader both men went to the messroom where there still was much excitement because of the promotion in rank. Captain Larson demanded that the crew [members] were to follow their leader and not to put the flag and the nation to shame. The doctors Brennecke, Barkow, Goeldel, and Heim, Captain Lorenzen, Second Officer Müller and Chief Engineer Heyneck immediately started packing their bags."[185]

Larsen confirmed the occurrences on arrival in Grytviken in his report, which he sent to the Norwegian king in mid-1913, who then had it confidentially forwarded to Kaiser Wilhelm II.[186] Everyone was alarmed when Larsen came on board. He tried to make it clear to Captain Lorenzen that he owed Filchner obedience, which Lorenzen strictly denied. Then Larsen addressed the crew, whom someone had apparently told that Filchner would have nothing to say anymore and that there

---

[185] Björvik: Schilderung einer Reise, pp. 57–66.
[186] Larsen, Carl Anton: (1913), Vertraulicher Bericht, Auswärtiges Amt, Berlin, AAV I B, Vol. 22, IIId 9021.

Fig. 41 After the expedition in South Georgia. *Back row (left)*: Lorenzen; *front row, from left to right*: Heim, Müller, and Barkow. (Joester, Erich (Private possession))

would be no money left to pay them after returning home. Larsen assured them that there were sufficient monetary funds and that he would personally guarantee their wages. After he had explained to them the consequences of mutiny against the expedition leader on board a ship, he advised them to apologize to Filchner.

When he spoke to Lorenzen again afterward, his group urged to move over to the *Harpune* immediately. Before they left Grytviken, Larsen impressed upon them the need to keep quiet about the troubles on board. Otherwise, he and the justice of the peace at Grytviken would give testimony that Lorenzen and some of the doctors had committed mutiny together with the crew. Only if they remained silent would he do nothing about it. The expedition's official stopover in Grytviken entailed that all participants received their polar bonus payments, which would be forfeited in the event of premature dismissal.[187]

To everyone's surprise in Germany, a message was received that Filchner's Antarctic expedition had already returned to Buenos Aires as early as January 7, 1913, although a further advancement on sea or land would have been possible at this time in the southern summer.[188]

---

[187] Discharge prior to the regular termination of a duty period is regulated by Article 57 of the Regulations for Seamen (Seemannsordnung): "5. if the voyage for which the seaman was hired cannot be embarked on or continued either due to warfare, embargo or blockade or a ban on import or export or because of any other incidental circumstance involving the ship or [its] cargo." (Knitschky: Die Seegesetzgebung, p. 168).

[188] Wichmann, Hugo: Südpolargebiete, in: Petermanns Geographische Mitteilungen 59 (1913). p. 30.

In late January, Filchner and his friends König and Przybyllok arrived in Berlin, whereas Brennecke, Barkow, Heim, Goeldel, Lorenzen, Müller, Heyneck, and nearly half of the crew returned home as late as mid-February and Kling remained on the *Deutschland* in Argentina.[189] Sometime later, *Geographische Zeitschrift* reported that Filchner had distanced himself from the idea of continuing the Antarctic expedition and that the *Deutschland* would now be called back home for the final disbandment of the expedition.[190]

In Berlin, Filchner did everything in his power to resolve the differences among the expedition members through the honorary council of his officer corps. In Wilhelmshaven, Filchner's proceedings against the ship's officers were in vain and he had to consent to a settlement, while the scientists' proceedings against Filchner failed in Munich.[191] Officially, this settled all of the altercations and insults that had occurred during the expedition; however, no one abided by the ruling of the honorary council. To restore his honor, Filchner challenged Goeldel to a duel with pistols; the latter declined and apologized in writing instead.[192]

Another consequence of the disputes was that Filchner's friend Przybyllok did not want to publish his results together with those of the other scientists in a comprehensive piece of work, and the other remaining scientists refused to publish their results together with Filchner. This resulted in numerous single publications appearing in various science journals instead of the initially planned collective publication in *Zeitschrift der Gesellschaft für Erdkunde zu Berlin*.[193]

According to Barkow, Filchner had "absolutely failed as an expedition leader and especially as a human being."[194] It is a fact that Filchner had completely underestimated life on board a ship with naval officers and a captain as the highest authority, particularly since he himself, a mere employee of an association, held no position of authority whatsoever on board. He hardly got along with the scientists, because, as a military man, he relied on command structures, whereas the scientists would accept nothing but a culture of discussion on equal terms. For Filchner, the expedition was over in Grytviken.[195] He "had had enough of 'Antarctic matters' for the time being."[196]

After returning home, König bought the *Deutschland* from the Association of the German Antarctic Expedition in order to continue Filchner's research work in Antarctica and further explore the sea arm between the Weddell Sea and the Ross Sea on an Austrian expedition, with the ship being renamed *Österreich*.[197] However, because of

---

[189] Drygalski, Erich von: Geographische Nachrichten, in: Mitteilungen der Geographischen Gesellschaft in München (1913), pp. 54–57.

[190] Anonymous: Süd-Polargegenden, in: Geographische Zeitschrift 19 (1913), p. 290.

[191] Penck, Albrecht: June 9, 1913, Brief an Herzog Ernst von Altenburg. Thüringisches Staatsarchiv Altenburg, Haus- und Privatarchiv der Herzöge von Sachsen-Altenburg, No. 226 h.

[192] Filchner: Feststellungen: pp. 49–50.

[193] Baschin, Otto: Note without title, in: Zeitschrift der Gesellschaft für Erdkunde zu Berlin 45 (1911), p. 497.

[194] Barkow: Diary, p. 216.

[195] Filchner: Ein Forscherleben, p. 134.

[196] Ibidem, p. 136.

[197] Anonymous: Süd-Polargegenden, p. 409; Filchner: Ein Forscherleben, p. 143.

the outbreak of World War I, the expedition could no longer leave the harbor of Trieste. The *Österreich* was sold to Rijeka in 1917 and operated as the *San Rocco* under the Italian flag until she was sunk at Korcula in 1926.[198]

The Scientific Results and the Aftermath

Despite the fact that Filchner's Antarctic expedition ended in a human disaster, it produced excellent results. Brennecke's oceanographic studies showed that there was a quadruple stratification of currents in the Southern Atlantic, moving southward from the equator, similar to Drygalski's discoveries in the Indian Ocean.[199] The meteorologist Erich Barkow did pioneering work during the drift in the ice. With 135 kite and balloon ascents up to an altitude of 2750 meters and a total of 120 pilot balloon ascents, rising up to 17,200 meters into the stratosphere, he continuously investigated the meteorological conditions over the Weddell Sea.[200] The geologist Heim came to the conclusion that the Ross Sea and the Weddell Sea might be connected by a subglacial trench.[201] Filchner's own descriptions of the break-off of the huge ice plate were most singular. Only much later did satellites make it possible to observe and monitor further break-offs of gigantic ice floes.[202]

Before World War I began, Roald Amundsen invited Filchner to take part in a new Arctic expedition, for which Filchner received pilot training at Johannisthal Airbase near Berlin and was trained as a cameraman by the company of Pathé Frères and Universum Film AG (UFA).[203] Amundsen even came to Berlin to visit AGO Flugzeugwerke in Johannisthal (Fig. 42). However, the expedition did not take place, because of the outbreak of war.

As a result of his adventurous expedition, Filchner could have become a German polar hero, but the premature end of the expedition, the separate homecoming of the participants, the missing home port reception festivities, the subsequent report on the

---

[198] Rack: Sozialhistorische Studie, p. 93.

[199] Brennecke, Wilhelm: Die ozeanographischen Arbeiten der Deutschen Antarktischen Expedition 1911–1912, in: Archiv der Deutschen Seewarte, 39 (1921), pp. 1–192; Lüdecke, Cornelia: Diverging Currents—Depicting Southern Ocean Currents in the Early 20th Century, in: Keith R. Benson/Helen M. Rozwadowski (ed.): Extremes: Oceanography's Adventures at the Poles, Sagamore Beach, pp. 71–105.

[200] Barkow, Erich: Vorläufiger Bericht über die meteorologischen Beobachtungen der Deutschen Antarktischen Expedition 1911/12, in: Veröffentlichungen des Preußischen Meteorologischen Instituts Berlin, No. 265, Abhandlungen Vol. 4 (1913) 11, pp. 7–8; Barkow, Erich: Die Ergebnisse der meteorologischen Beobachtungen der DAE 1911/12 (posthumous), in: Veröffentlichungen des Preußischen Meteorologischen Instituts, No. 325, Abhandlungen Vol. 7 (1924) 6.

[201] Heim, Fritz: Wissenschaftliche Ergebnisse der II. deutschen Südpolarexpedition, in: Mitteilungen der Geographischen Gesellschaft in München 1914, p. 509.

[202] Filchner: Zum sechsten Erdteil, pp. 245–272.

[203] Filchner: Ein Forscherleben, p. 141, 143; Lüdecke, Cornelia: Roald Amundsen. Ein biografisches Portrait, Freiburg 2011, pp. 113–114.

Fig. 42 Roald Amundsen visiting AGO Flugzeugwerke in Berlin-Johannisthal. *From left to right*: the directors Hermann von Frémery and Elisabeth Woerner (Mrs. Frémery), and Amundsen. (Filchner Archive, Munich)

experiences of the adventurers, and the usual eulogies failed to deliver headlines in the daily newspapers. In comparison with his later expeditions to Tibet and China, the memory of Filchner's polar adventures was relegated to the background.[204]

Excursion: Continuity after World War I—The Foundation of the German Society of Polar Research [Gesellschaft für Polarforschung]

When no expensive Antarctic expeditions could be sponsored in the aftermath of World War I, researchers began to focus on the Artic region again. At first, they carried out private expeditions—for example, that of Max Grotewahl, who, in 1925, set out to Spitsbergen with three companions and was picked up by the *Zieten* with Alfred Ritscher as the polar expert on board (Fig. 43).[205]

---

[204] Filchner, Wilhelm: Om mani padme hum. Meine China und Tibetexpedition, Reprint. Wiesbaden 2013.
[205] Lüdecke, Cornelia: Zum 100. Geburtstag von Max Grotewahl (1894–1958), Gründer des Archiv für Polarforschung, in: Polarforschung 65 (1995), pp: 93–105.

Fig. 43 Participants in the Spitsbergen expedition and the officers on board the *Zieten*, photographed in late August 1925. *From left to right, front row*: Max Grotewahl (expedition leader), the *Zieten* captain Alfred Ritscher (polar expert), and Rudolf Jupitz (responsible for biology); *back row*: two officers, Fritz Biller (responsible for glaciology), three officers, and Walter Ankersen (cameraman). (Archiv für deutsche Polarforschung, Alfred-Wegener-Institut, Bremerhaven, Grotewahl estate)

However, Grotewahl had to deal with huge problems while making preparations for his expedition, because there was no central institution with a comprehensive and specialized library and map collection, where he also might have got information on equipment and expedition technology. To overcome this setback, he founded the Polar Research Archive [Archiv für Polarforschung] in his hometown of Kiel on July 1, 1926, hoping that a polar research institute [Institut für Polarforschung] would develop from it someday.[206] For financial support, the Association for the Promotion of the Polar Research Archive [Vereinigung zur Förderung des Archivs für Polarforschung] was founded in 1927 in order to make every effort to support the recommencement of polar research activities. In 1931, the association already had 108 members and the first edition of *Polarforschung* was published.[207] The archive became known in the press because Grotewahl, as the "director of the polar research archive," commented on polar topics and published reports on expeditions in the newspapers (Fig. 44).

Twenty-five years after its foundation, the library of the archive contained 6000 volumes and reprints, 350 maps, 49 German journals, and 127 foreign journals, which were

---

[206] Scholz, Arnulf: 5 Jahre »Archiv für Polarforschung«, in: Polarforschung 2 (1932), p. 2; Grotewahl, Max: Prof. Dr. Max Robitzsch, in: Polarforschung 22 (1952), p. 145.

[207] Mitteilungen für die Vereinigung zur Förderung des Archivs für Polarforschung (1931), p. 3.

Fig. 44 Max Grotewahl at the Polar Research Archive (in Kiel) in late September 1930. (Archiv für deutsche Polarforschung, Alfred-Wegener-Institut, Bremerhaven, Grotewahl estate)

available by way of exchange with other institutions.[208] Grotewahl even succeeded in releasing at least one issue of each annual volume of the journal, although with some delay. Apart from the participants in Wegener's expedition to Greenland (1930–1931), Alfred Ritscher (the leader of the third German Antarctic Expedition (1938–1939)) and his second wife Ilse (née Uhlmann) were later also active in the promotional association. On the occasion of the archive's 25th anniversary in 1951, the first International Polar Conference [Erste Internationale Polartagung] took place in Kiel, at which Ritscher was voted in as its first president.[209]

After Grotewahl had passed away in 1958, the promotional association was renamed the German Society for Polar Research [Deutsche Gesellschaft für Polarforschnug e.V.] at the Second Polar Conference in Holzminden, and its presidency was held by Alfred Ritscher for as long as he lived. In the course of several decades, the society developed into an internationally acknowledged professional association, which nowadays holds an international meeting in Germany, Austria, or Switzerland every two and a half years.

---

[208] Tiedemann, Karl-Heinz/Ruthe, Kurt: 25 Jahre Archiv für Polarforschung, in: Polarforschung 21 (1951), p. 82.

[209] Weiken. Karl: Prof. Dr. Bernhard Brockamps Verdienste um die deutsche Polarforschung und um die Deutsche Gesellschaft für Polarforschung, in: Polarforschung, 38 (1968), p. 192.

The Discovery of Neu-Schwabenland: The Third German South Polar Expedition (1938–1939)

The "Fat Gap", Whaling, and German Possession Claims in Antarctica

After World War I, domestic agriculture in Germany was unable to satisfy the increased demands for dietary and industrial fats with its own produce. In 1932, the consumption of margarine and artificial dietary fats was covered by foreign imports amounting to 524,000 tons, compared with 2000 tons of domestic production (0.6%).[1] Whale oil (train oil) had already been secretly used as a basic ingredient in margarine during World War I and was increasingly needed as a raw material for lubricants and soaps.[2] The growing demand produced a so-called "fat gap", which cost the country important foreign currency—an expense that needed to be avoided after the National Socialists seized power.[3] In line with the "fat plan," consumption was to be gradually switched to domestic fats. As a first step, the need to produce industrial greases, soap, and dietary fats was to be satisfied by increased domestic oilseed production. However, whale oil had to be imported as long as the agricultural economy was unable to achieve this objective. This soon resulted in an acute lack of foreign currency. Import quotas on train oils were therefore put in place to protect domestic production.[4]

---

[1] Backe, Herbert: Um die Nahrungsfreiheit Europas. Weltwirtschaft oder Großraum. Leipzig 1942, p. 287; see also Lüdecke, Cornelia: In geheimer Mission zur Antarktis. Die dritte Deutsche Antarktisexpedition 1938/39 und der Plan einer territorialen Festsetzung zur Sicherung des Walfangs, in: Deutsches Schiffahrtsarchiv 26 (2003), pp. 75–100.

[2] Winterhoff, Edmund: Walfang in der Antarktis, Oldenburg 1974, pp. 54–55, 58.

[3] Hugo, Otto: Deutscher Walfang in der Antarktis, Oldenburg i. O. 1939: pp. 40–51; Winterhoff: Walfang, p. 70; see also Scholl, Lars-Uwe: German Whaling in the 1930s, in: Fischer, Lewis R./Norvik, Helge W./Minchinton, Walter E. (ed.): Shipping and trade in the Northern Seas 1600–1939. Yearbook of the Association for the History of the Northern Seas 1988, pp. 106–107.

[4] Knaurs Konversationslexikon, Berlin 1934, column 410, keyword "Fettwirtschaft."

© Springer Nature Switzerland AG 2021
C. Lüdecke, *Germans in the Antarctic*,
https://doi.org/10.1007/978-3-030-40924-1_3

In 1933, Helmuth Wohlthat became the director of the newly established Reich Agency for Dairy Products, Oils and Fats [Reichsstelle für Milcherzeugnisse, Öle und Fette] in Berlin and, as such, had a significant role to play in the later Antarctic expedition. As early as September 1934, he was appointed as the departmental head for the "New Plan," which was developed by Hjalmar Schacht at the Reich Ministry of Economics [Reichswirtschaftsministerium] and intended to regulate imports by following the principle of "foreign trade without foreign currency."[5] In January 1935, Wohlthat finally became the director of the Reich Foreign Currency Office [Reichsdevisenstelle] at the Reich Ministry of Economics.[6] The fat gap and foreign currency management—both topics of focus during Wohlthat's career—were closely related and had a considerable influence on the resumption of Germany's South Polar research.

After Germany became the second-largest purchaser of Norwegian harvests of whale oil, importing 150,000–200,000 tons each year, the double price increase for a ton of train oil from 10 pounds sterling (200 Reich marks) for the season of 1934–1935 to 20 pounds sterling (400 Reich marks) for the season of 1935–1936 had a substantial impact.[7] Thus, foreign currency payments to Norway became the highest individual expense item.[8] These circumstances promoted the building of a German whaling fleet in order to achieve independence from Norwegian whale oil imports.[9]

In this context stands the foundation of Walter Rau Walfang AG by Walter Rau, the proprietor of the largest trust-free factory. He commissioned the construction of the most modern fleet of whaling ships, including the *Walter Rau* whaling factory mothership and eight whale catcher boats named *Rau I* to *Rau VIII* (Figs. 1 and 2).[10]

Since lessons had been learned from the ruthless exploitation by whaling in the Arctic, a total of 19 states adopted the Geneva International Convention for the Regulation of Whaling early in 1936.[11] Germany signed the agreement but never ratified it.

---

[5] See also Wohlthat, Helmuth: Neue Entwicklungsmöglichkeiten des deutschen Verrechnungsverkehrs, in: Staatenwirtschaft, ständige Beilage zur Zeitschrift für Geopolitik 3 (1939) 4/5, pp. 701–706; Pentzlin, Heinz: Hjalmar Schacht. Leben und Wirken einer umstrittenen Persönlichkeit, Berlin 1980, p. 216; Winterhoff: Walfang, pp. 71–72.

[6] Winterhoff: Walfang, p. 72; Pentzlin: Hjalmar Schacht, p. 216.

[7] Wohlthat, Helmuth: Walöl im Weltmarkt. Der Vierjahresplan 3 (1939) 11, p. 730; Peters, Nicolaus: Über Hochseewalfang und Tierleben im Südlichen Eismeer, in: Der Fischmarkt 1937 (7/8), p. 19; Herrmann, Ernst: Deutsche Forscher im Südpolarmeer, Berlin 1941: p. 176.

[8] Wohlthat: Walöl im Weltmarkt, p. 731.

[9] Winterhoff: Walfang, pp. 71–73.

[10] Peters, Nicolaus: Kurze Geschichte des Walfangs von den ältesten Zeiten bis heute, in: Peters, Nicolaus (ed.): Der neue deutsche Walfang, Hamburg 1938, pp. 20–21; Hugo: Deutscher Walfang, pp. 40–72. The first whaling voyage of the *Walter Rau* is described in: Spengemann, Herbert: Auf Walfang in der Antarktis, Bühl-Baden 1939.

[11] Ahlbrecht, Bernhard: Internationale Walfangabkommen, deutsches Walfanggesetz und Reichsförderung, in: Peters, Nicolaus (ed.): Der neue deutsche Walfang, Hamburg 1938, pp. 28–30.

Fig. 1 The whaling factory mothership *Walter Rau* and two whale catcher boats. (Spengemann, Herbert: Auf Walfang in der Antarktis, Bühl-Baden 1939, p. 65)

Fig. 2 The whale catcher boats *Rau I* to *Rau VIII* in the harbor at Wesermünde in 1938. (Spengemann, Herbert: Auf Walfang in der Antarktis, Bühl-Baden 1939, p. 16)

In order to take action against the aggravating raw material and fuel crisis in the long term, the first Four-Year Plan, pursuing the goal of economic autarky, was proclaimed at the Nationalsozialistische Deutsche Arbeiterpartei (NSDAP) [National Socialist German Workers' Party] Reich Party Congress on September 9, 1936. However, the plan was used to pursue military rearmament within the next four years.[12] Hermann Göring, the commander-in-chief of the German Air Force [Luftwaffe], was appointed as the commissioner of the Four-Year Plan [Beauftragter für den Vierjahresplan] and endowed with comprehensive authority to make preparations for war. Self-sufficiency was to be achieved "regardless of the costs." Whale oil was a significant factor; after all, it was also one of the most essential raw materials in the production of glycerin, which was needed to manufacture explosives.[13]

German whaling operations began in the season of 1936–1937 under the command of Captain Otto Kraul, who was sailing on the *Jan Wellem*, a former Hapag-Lloyd steamship recently rebuilt as a whaling factory ship.[14] Together with at first six and later eight whale catcher boats, named *Treff I* to *Treff VIII*, they constituted the first German whaling fleet, which was commissioned by the Henkel company and operated by the First German Whaling Company, Ltd. [Erste Deutsche Walfang-Gesellschaft] in order to supply the fat raw materials urgently needed for the company's various products.

In addition, the Margarine Raw Material Procurement Company [Margarine-Rohstoff-Beschaffungsgesellschaft] in Berlin chartered the two Norwegian whaling fleets *C.A. Larsen* and *Skytteren*, for whose operation the Hamburg Whaling Counting House [Hamburger Walfang-Kontor] was founded in June 1937. In the same year, the latter also added the purchased *Sydis*, under the name *Südmeer*, to their hunting vessels.

The Unilever Corporation had the mothership *Unitas* built and deployed it, together with eight whale catcher boats, in the whaling season of 1937–1938. Hence, it was not a coincidence that the whaling enterprises were referred to in this context as "Germany's colonial possessions in Antarctica."[15]

The production of whale oil in this season covered nearly one half of the quantity needed for the manufacture of margarine (Figs. 3 and 4).[16] The sea was even said to be the "new German fat colony," as the German fleet produced 95,000 tons of whale oil; thus, only 110,000 tons still had to be imported.[17]

---

[12] See also Wohlthat: Neue Entwicklungsmöglichkeiten, p. 702; Petzina, Dietmar: Die deutsche Wirtschaft in der Zwischenkriegszeit, Wiesbaden 1977, pp. 124–139, Broszat/Frei: Ploetz, pp. 61–61, 111–112.

[13] Knaurs Konversationslexikon, column 320, 410, 513.

[14] Frank, Wolfgang: Der wiedererstandene deutsche Walfang. Dargestellt an der Entwicklungsgeschichte der ersten deutschen Walfang-Gesellschaft in Verbindung mit einem Reisebericht über die 2. »Jan-Wellem«-Expedition, Düsseldorf 1939; Hugo: Deutscher Walfang, pp. 69–76; Winterhoff: Walfang, pp. 77–78, 85–88, 117–125.

[15] Wegener, Karl August: Die deutsche Kolonie in der Antarktis, in: Peters, Nicolaus (ed.): Der neue deutsche Walfang, Hamburg 1938, p. 5.

[16] Wohlthat: Neue Entwicklungsmöglichkeiten, p. 613.

[17] Lübke, Anton: Das deutsche Rohstoffwunder. Wandlungen der deutschen Rohstoffwirtschaft, Stuttgart 1943, pp. 447–448.

The Plan for a New German Antarctic Expedition

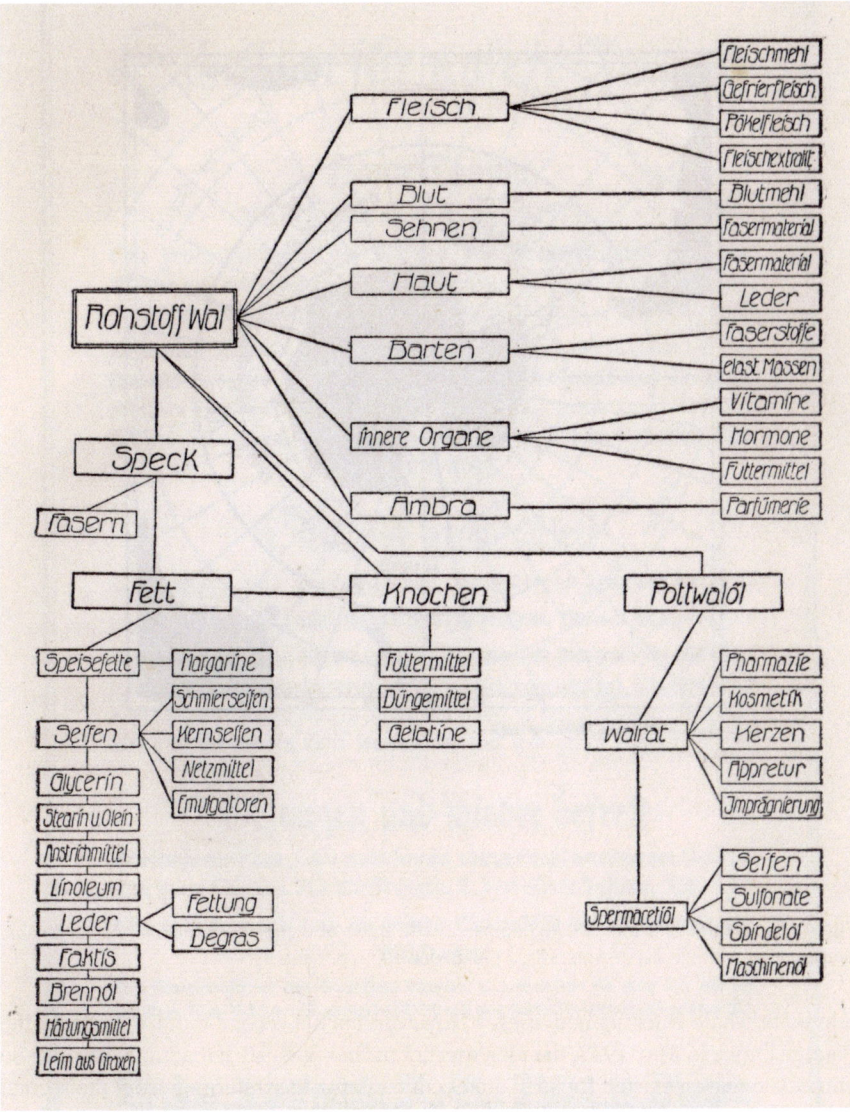

Fig. 3 Products of whale processing. (Spengemann, Herbert: Auf Walfang in der Antarktis, Bühl-Baden 1939, p. 103)

The Plan for a New German Antarctic Expedition

After the German whaling fleet had set out to the Antarctic seas in 1936, the German Foreign Office contemplated, for the first time, whether and how German territorial claims on the continent of Antarctica could be asserted to protect the economically

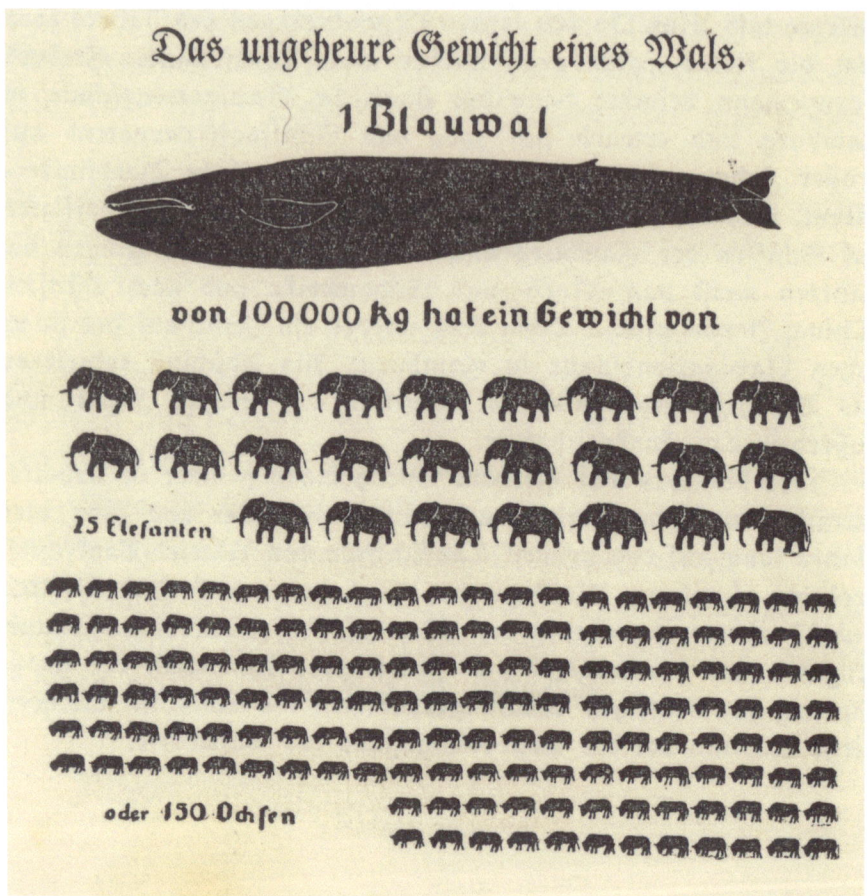

Fig. 4 The weight of a blue whale compared with those of elephants and oxen. (Spengemann, Herbert: Auf Walfang in der Antarktis, Bühl-Baden 1939, p. 100)

important whale-catching activities.[18] However, upon a second inquiry made by the Foreign Office in May 1937, the idea was not further pursued at that time, because no immediate reason existed for such a procedure. Only after returning from the second whaling season did Wohlthat, who became Minister for Special Affairs at the commissioner of the Four-Year Plan, submit to his superior, Hermann Göring, a plan for a German Antarctic expedition.[19] Since greater economic interest in the South Polar region had now become obvious, Göring revisited the proposal and commissioned

---

[18] Auswärtiges Amt: June 11, 1938, Anlage Protokoll der Sitzung vom June 11, 1938, No. R 13428. Leibniz-Institut für Länderkunde, Leipzig, Ritscher estate estate, File AA1, Abt. AA; Deutscher Kolonial-Dienst: issue of May 15, 1938.

[19] Ritscher, Alfred (ed.): Wissenschaftliche und fliegerische Ergebnisse der Deutschen Antarktischen Expedition 1938/39, Vol. 1, Leipzig 1942 p. IX, 2.

Wohlthat with its organization.[20] In a meeting that took place at the Foreign Office on June 11, 1937, and included the representatives of several ministries, it was determined that there was no discernible military interest in Antarctica connected with the Four-Year Plan. While some states were pursuing territorial claims, the territorial question was to be avoided at the forthcoming London Whaling Conference in order to reaffirm Germany's outwardly harmless intentions. The legal issues of occupation of Antarctic territory were examined by the Foreign Office, which stated that the current development revealed "clearly an impending monopolization of the Antarctic continent by the United States, England including her possessions, Norway and France, excluding all outsiders."[21] For the Foreign Office, the investigations that had been conducted had proved that there were no sufficient legal foundations for German claims on the basis of international law at the time. As a solution, it was therefore suggested that a station should be established on land in Antarctica as early as the next winter, which, under the cover of research activity, was to lay a foundation for later occupation, considering the whaling interests and the natural resources assumed to exist in Antarctica, as well as future flight connections. It would also have to be checked whether establishment of a special institute would be advisable to carry out such an expedition. On the part of the Ministry of Aviation [Reichsluftfahrtministerium], it even appeared desirable to "presently gain an airbase for future aviation which could instantly serve as a weather and radio station. In the future, the flight route from South America to Australia would undoubtedly go over Antarctica" (Fig. 5).[22]

It was quite easy to bring the Antarctic expedition into agreement with the Four-Year Plan, which was rearranged in summer with regard to the war economics mobilization, pursuing the objective of becoming absolutely independent of foreign raw material supplies.[23] The Ministry of Aviation believed that accelerated inclusion of Germany in the circle of powers interested in Antarctica was inevitable, and it took on the leadership responsibility until it could be handed over to the scientific leader.[24] It also wanted to declare its readiness for preparing and assuming leadership if the enterprise were to predominately appear as a scientific expedition. The first step was to establish an expedition office at German Lufthansa at Berlin Tempelhof.[25]

Early in August, the high command of the German War Navy [Kriegsmarine] (formerly known as the Imperial Naval Office) proposed that the captain and

---

[20] Auswärtiges Amt: June 3, 1938, Notiz to 1.Abtlk.Sk l.1981/38 geh., Leibniz-Institut für Länderkunde, Leipzig, Ritscher estate, File AA1, Abt. AA; Auswärtiges Amt, June 11, 1938.

[21] Ibidem.

[22] Ibidem.

[23] Petzina: Die deutsche Wirtschaft, pp. 128–129, 132.

[24] Oberkommando der Kriegsmarine: August 11, 1938, to B.No. 420/58 geh. BH VII Ang., Leibniz-Institut für Länderkunde, Leipzig, Ritscher estate, File Bh1, Abt. OKM; Reichsministerium für Ernährung und Landwirtschaft: June 29, 1938. II B 57201 II, Leibniz-Institut für Länderkunde, Leipzig, Ritscher estate, File AA1, Abt. AA.

[25] Ritscher, Alfred: August 31, 1938, Brief an Beauftragter für den Vierjahresplan, Minister zur besonderen Verwendung, Leibniz-Institut für Länderkunde, Leipzig, Ritscher estate, File Bh1, Abt. VJP.

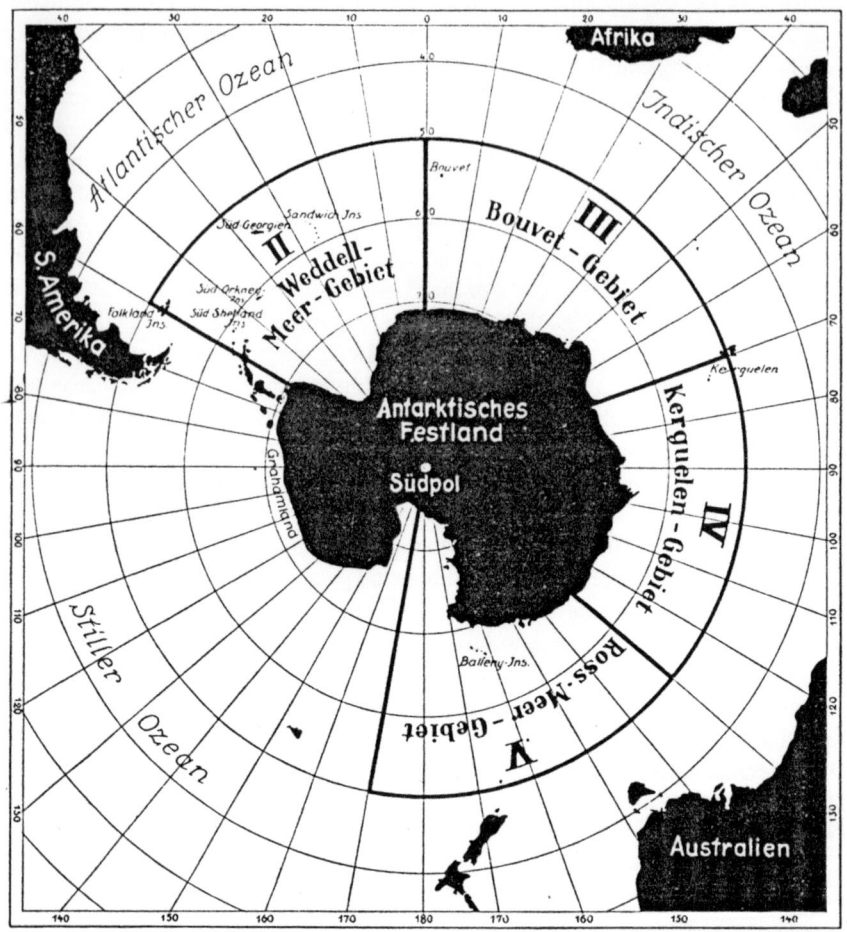

Fig. 5 Catching sectors II–V in Antarctic high sea whaling. (Peters, Nicolaus: Über Hochseewalfang und Tierleben im Südlichen Eismeer, in: Der Fischmarkt 1937 (7/8), p. 2)

aviation pilot Alfred Ritscher, who was working for the navy's Maritime Handbook Division [See-Handbuch-Werk], was to be the leader of the expedition.[26] As the captain of the ill-fated Schröder-Stranz expedition to Spitsbergen in 1912, Ritscher had abundant previous polar experience and had proved his worth organizing air travel in South America.[27] Early in October, Göring commanded him to travel to Antarctica in order to carry out a scientific flight expedition and appointed him as

---

[26] Oberkommando der Kriegsmarine: August 4, 1938, Zur Besprechung vom 15. Juli 1938, B.No. 420/38 geh. BH V, Leibniz-Institut für Länderkunde, Leipzig, Ritscher estate, File Bh1, Abt. OKM.

[27] Ritscher, Alfred: Wanderung in Spitzbergen im Winter 1912, in: Zeitschrift der Gesellschaft für Erdkunde zu Berlin 1916, pp. 16–34; Ritscher: Wissenschaftliche und fliegerische Ergebnisse, p. 3; Hartmann, Gertraude: 2014, 1926 Tagebuch Südamerika—Vorbereitungen »Luther-Flug«. aufbereitet und recherchiert für ihre Familie von Gertraude Hartmann, geborene Ritscher, unpublished, private possession Hartmann, Braunfels.

the "leader of the same expedition, effective as of September 1, 1938."[28] Wohlthat devised the organizational plan for the preparation and execution of the third German Antarctic Expedition (1938–1939) (DAE 38/39), while Ritscher was to take care of the practical aspects of its preparation and execution.[29]

Alfred Ritscher
Alfred Ritscher was born in Bad Lauterberg on May 23, 1879, as the second son of a medical doctor. His father died when he was barely two years old. From 1897 onward, he went to sea on sailing ships and in 1903 acquired a second mate's certificate of competency.[30] He then completed one year of voluntary military service as a reserve officer candidate in the Imperial German Navy. In late 1904, he was employed by the Hamburg South America Line. Finally, in February 1907, he obtained his captain's license. As a result of his regular participation in navy duty training, he was incidentally promoted to the rank of a second lieutenant and then a first lieutenant of the reserve in the German Navy. In November 1911, he found employment in the Maritime Handbook Division of the Nautical Department of the Imperial Naval Office [Reichsmarineamt] in Berlin. In July 1913, he was furloughed to take part in the Schröder-Stranz expedition to Spitsbergen as the captain of the expedition ship *Herzog Ernst* (Fig. 6).[31]

Schröder-Stranz had himself and three of his comrades dropped off on the northern coast of North East Land [Nordaustlandet] in order for them to be picked up again on the southwestern coast after they had traversed the island; the option of overwintering was part of the plan. The *Herzog Ernst* happened to get trapped in ice in Treurenburg Bay on Spitsbergen's northern coast. Therefore, the expedition members decided to fight their way through to the

(continued)

---

[28] Beauftragter für den Vierjahresplan: October 5, 1938, ST.M.Dev. 1241 g.Rs., Leibniz-Institut für Länderkunde, Leipzig, Ritscher estate, File Bh1, Abt. VJP.
[29] Wohlthat, Helmuth: October 10, 1938, W XVII/91 g.Rs., Leibniz-Institut für Länderkunde, Leipzig, Ritscher estate, File Bh1, Abt. VJP; Ritscher: August 31, 1938.
[30] Georgi, Johannes: Polarforscher Kapitän Alfred Ritscher, in: Polarforschung 32 (1962) 1/2, pp. 125–127; Stocks, Theodor: In Memoriam Alfred Ritscher 1879–1963, in: Deutsche Hydrographische Zeitschrift 16 (1963) 3, pp. 87–92; Hartmann, Gertraude: 2007, Auszug aus der Ahnenforschung von Alfred Ritscher, überarbeitet und ergänzt von seiner Tochter Gertraude Hartmann, geborere Ritscher, unpublished, private possession Hartmann, Braunfels; Lüdecke, Cornelia: Die deutsche Polarforschung seit der Jahrhundertwende und der Einfluß Erich von Drygalskis, Dissertation, in: Berichte zur Polarforschung 158 (1995), Anhang A8–A9.
[31] Rüdiger, Hermann: Die Sorge-Bai. Aus den Schicksalstagen der Schröder-Stranz-Expedition, Berlin 1913; Ritscher: Wanderung in Spitzbergen.

Fig. 6 The expedition members on board the *Herzog Ernst* in 1912. *From left to right, front row*: Sandleben, Schröder-Stranz, [unknown], Ritscher, and Rave; *second row*: Moeser, Rüdiger, Detmers, and Mayr. (Leibniz-Institut für Länderkunde, Leipzig, Ritscher estate)

coal settlement of Advent City to avoid overwintering on the island. On the way, however, they split up into several groups. In the end, after an adventurous march through the polar night, Ritscher succeeded in reaching the coal settlement, accompanied only by his dog, and could seek help for the scattered groups. However, no trace could be found of Schröder-Stranz, his companions, and three other expedition members, while an eighth member fell ill and died during the overwintering.

After the outbreak of World War I, Ritscher was first an aircraft pilot and then a flight squad leader at the Navy Command in Flanders.[32] He got his pilot's license at Johannisthal Airbase near Berlin on February 10, 1915. On

(continued)

---

[32] Georgi: Polarforscher Kapitän Alfred Ritscher; Stocks: In Memoriam; Hartmann: Auszug aus der Ahnenforschung.

August 20, 1915, he married the artist Susanne Loewenthal (1886–1975). Later, as a commander of field aviators [Landflieger], he was transferred to a navy base near Oldenburg. After the war, he once again worked for the Maritime Handbook Division until he was laid off because of inflation in 1919. Only in July 1924 could he resume his work there.

Since the opportunities for promotion at the Imperial Naval Office were limited, Ritscher transferred to the newly founded German Lufthansa as the head of the Aircraft Navigation Department [Abteilung Flugzeugnavigation] in April 1925.

When the former Reich Chancellor Hans Luther traveled to South America as an administrative advisor for the German Reich Railway [Deutsche Reichsbahn] in order to rebuild the economic cooperation with the German population living there, which had been suspended because of World War I, Lufthansa was commissioned to establish a flight route between Buenos Aires (Argentina) and Natal (Brazil) because of unsatisfactory traffic conditions.[33] This secret commission went by the name of the Iguazu Flight. Ritscher traveled to Rio de Janeiro to inspect all landing and takeoff possibilities of such a coastal journey and to install a temporary weather information and flight control service at a few coastal radio stations for the duration of the flight. In addition, he made sure there was enough fuel to fill the tanks of the two Dornier 8-t-Wal seaplanes at their anchoring sites. The nearly 2400-kilometer-long flight (during November 19–27, 1926) was completed without any complications.

In 1929, the high command of the German War Navy called on Ritscher to rebuild the new navy aviation sector as part of the Reich Association of the German Aviation Industry [Reichsverband der Deutschen Luftfahrtindustrie], where he was responsible for development of modern air navigation devices.[34]

After the National Socialists seized power in 1933, the Reich Association of the German Aviation Industry was taken over by the Aviation Ministry [Luftfahrtministerium] and Ritscher returned to his former work at the Maritime Handbook Division. In September 1933, he divorced his Jewish wife Susanne to avoid career disadvantages. On May 1, 1934, he attained civil servant status and was appointed as a senior executive officer [Regierungsrat], whereas Susanne Ritscher escaped the Holocaust only because she feigned suicide in 1944 and went into hiding in various places until World War II was over.[35]

(continued)

---

[33] Hartmann: 1926 Tagebuch.

[34] Georgi: Polarforscher Kapitän Alfred Ritscher; Stocks: In Memoriam; Hartmann: Auszug aus der Ahnenforschung.

[35] Pietsch, Jani: »Ich besaß einen Garten in Schöneiche bei Berlin«: Das verwaltete Verschwinden jüdischer Nachbarn und ihre schwierige Rückkehr, Frankfurt am Main 2006, pp. 123–135.

Fig. 7 Alfred Ritscher in 1942. (Ritscher, Alfred (ed.): Wissenschaftliche und fliegerische Ergebnisse der Deutschen Antarktischen Expedition 1938/39, Leipzig 1942, p. 5)

In August 1938, the Office of the Four-Year Plan gave Ritscher a government mandate, which was approved by the German War Navy, to prepare for, and assume leadership of, an expedition to Antarctica. The third German Antarctic Expedition (1938–1939), using the catapult ship *Schwabenland* and the Dornier 10-t-Wal flying boats *Boreas* and *Passat*, lasted from December 17, 1938, until April 12, 1939.

From January 15, 1941, onward, Ritscher interrupted his work at the Office of the German War Navy and was deployed in World War II. Late in March, he married Ilse Uhlmann, the secretary for his Antarctic expedition. He was a lieutenant commander of the reserve and a commandant of a minesweeper in the English Channel and the Bay of Biscay. Other deployment locations were Croatia and North Italy, where he was taken prisoner by the English and released as early as late August 1945 (Fig. 7).

In 1951, Ritscher was elected as the chairman of the German Society of Polar Research.

(continued)

> The second volume of the results of the third German Antarctic expedition (1938–1939) appeared only in late 1958.
> On March 30, 1963, Ritscher died in Hamburg and was interred in Bad Lauterberg.

Preparations for the *Schwabenland* Expedition

On September 3, 1938, Hermann Göring defined the assignments of the Antarctic expedition for German Lufthansa in the form of Secret Reich Matter.[36] Within a short summer campaign, the hinterland of the hitherto unknown coastal area east of the Weddell Sea was to be explored between 20°W and 20°E in order to lay a foundation for sovereignty by dropping flags for the National Socialist Reich's later occupation to safeguard German whaling activities (Fig. 8). Meteorological, oceanographic, and geomagnetic observations were to provide the information needed for the expedition but also served the purpose of concealing the enterprise from the eyes of the world. In addition, the expedition was to be kept secret for as long as possible. After the return to Germany, proposals were to be made for a further expedition into the area between Graham Land and the Ross Sea in the course of the southern summer (1939–1940).

After North German Lloyd and German Lufthansa had declined to take over the official operation of the expedition ship, Ritscher recommended the Kaiser Wilhelm Society for the Advancement of Science [Kaiser-Wilhelm-Gesellschaft zur Förderung der Wissenschaften] (now known as the Max Planck Society) in Berlin as its operator—a choice that was greatly appreciated by Wohlthat, as the concealment of the expedition would thus be best assured on the scientific side.[37] Originally, the catapult ship *Westfalen*, on duty in the South Atlantic, had been intended to be used as the expedition ship, but ultimately the (actually less suitable) *Schwabenland* was deployed for reasons of timeliness.[38]

---

[36] Beauftragter für den Vierjahresplan: September 3, 1938, Göring an Deutsche Lufthansa, ST.M.Dev. 1075 g.Rs., Leibniz-Institut für Länderkunde, Leipzig, Ritscher estate, File Bh1, Abt. VJP.
[37] Wohlthat, Helmuth: October 19, 1938, W XVII/106, Leibniz-Institut für Länderkunde, Leipzig, Ritscher estate, File Bh1, Abt. VJP.
[38] Ritscher: Wissenschaftliche und fliegerische Ergebnisse, pp. 15–21.

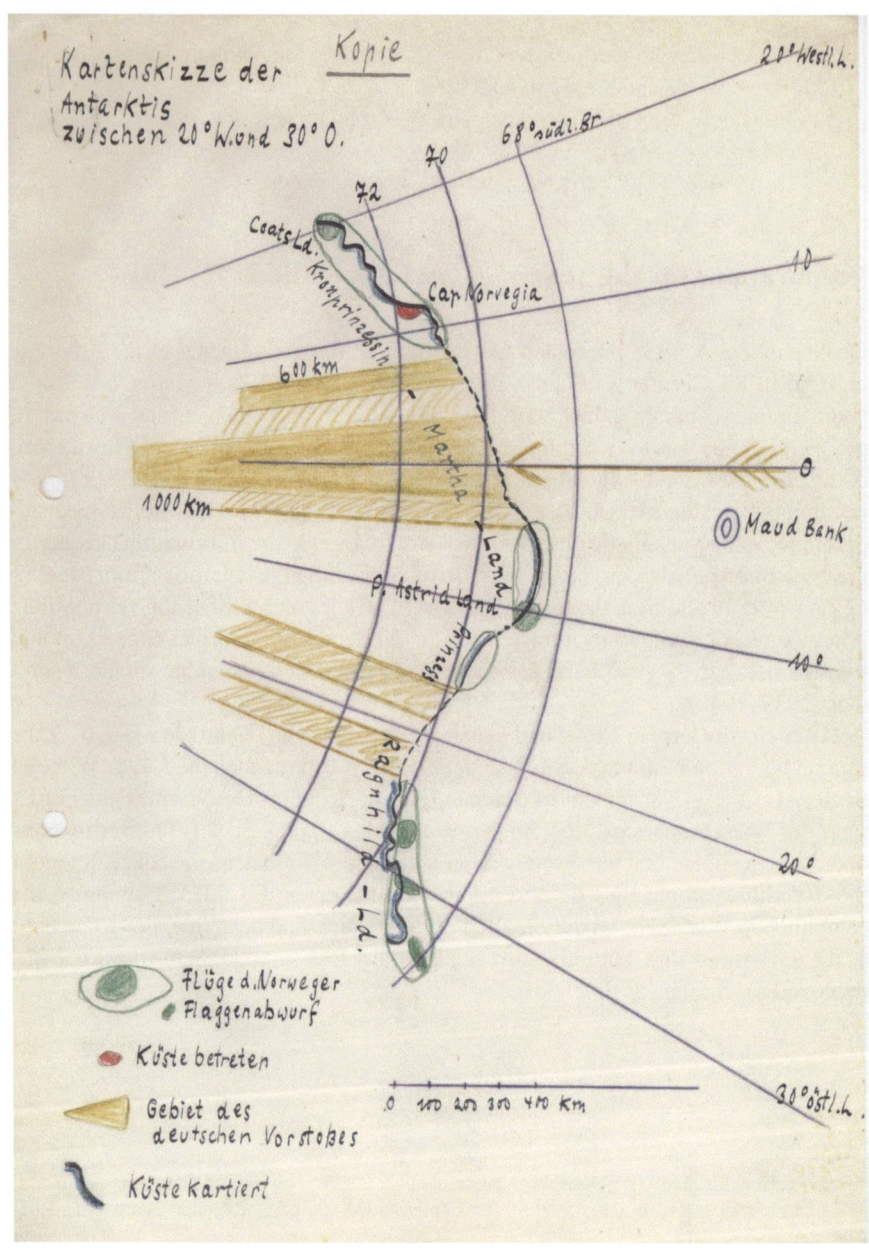

Fig. 8 Ernst Herrmann's drawing of the geographic work possible in Antarctica, done in 1938. (Leibniz-Institut für Länderkunde, Leipzig, Ritscher estate)

The *Schwabenland* and the Flying Boats *Boreas* and *Passat*

The motorship *Schwarzenfels* (built in 1925) was converted for German Lufthansa into the catapult ship *Schwabenland* for her deployment as an airbase in transatlantic aviation in 1934.[39] In particular, she received a catapult with an acceleration track of 31.6 meters for launching and a crane for picking up flying boats, which could then be refueled and maintained on board. The *Schwabenland* was overhauled again to meet the special requirements of the polar expedition.

Technical details of the ship:

> Length of the solid upper deck: 142.7 meters (Fig. 9)
> Maximum width: 18.4 meters
> Molded depth of the hull up to the main deck: 10.65 meters
> Gross tonnage: 8488 gross register tons
> Tanks for a total of 1785 tons of fuel for an action radius of 24,000 nautical miles (44,448 kilometers)
> Two four-stroke diesel main engines, each with 1800 horsepower
> Four addition auxiliary diesel engines
> Average speed: 10¾–11 nautical miles per hour (approximately 19 kilometers per hour)

Special equipment on the ship:

> Catapult model Heinkel K7 for maximum launch weights of 14,000 kilograms (Fig. 10)
> Laterally applied park station for the second flying boat

Fig. 9 The catapult ship *Schwabenland*. (Leibniz-Institut für Länderkunde, Leipzig, Herrmann estate)

(continued)

---

[39] Ibidem; Buddenbrock, Friedrich Frhr. von: »Atlantico« »Pazifico«. Lehrjahre des überseeischen Luftverkehrs, Düsseldorf, 1965: Anhang.

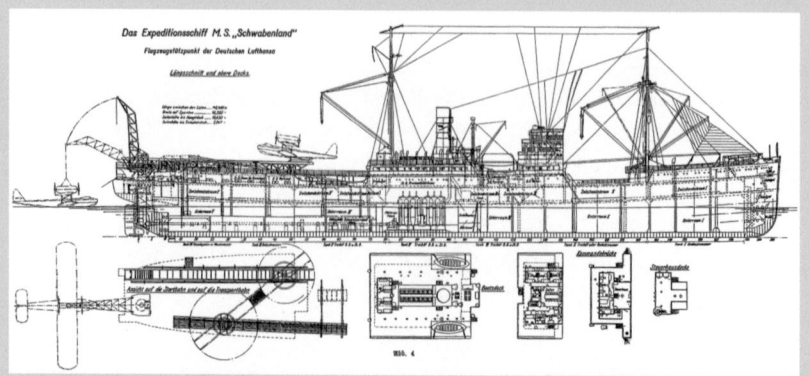

Fig. 10 *Above:* A longitudinal section; *below:* the superstructure of the *Schwabenland*. (Ritscher, Alfred (ed.): Wissenschaftliche und fliegerische Ergebnisse der Deutschen Antarktischen Expedition 1938/39, Leipzig 1942, p. 16)

> Crane with a tiltable boom to pick up flying boats; lifting capacity 12,000 kilograms
> Landing canvas to facilitate the pickup of flying boats
> Radio installation and tracking device for long-wave, medium-wave, short-wave, and ultrashort-wave radio for distances of up to 600 kilometers
> Anschütz gyroscopic compass system
> Echo-sounding control center to determine sea depths with a universal echo sounder manufactured by Elektro-Akustik, consisting of one shallow sounder (magnetostriction sounder, 30 kilohertz) and a deepsea sounder (echo sounder, 3 kilohertz), as well as one Atlas deepsea sounder (echo sounder, 3 kilohertz)

Two flying boats of the type Dornier 10-t-Wal from German Lufthansa, designed for South Atlantic postal transport, were made available for aerial photogrammetric images. They were equipped with huge additional tanks for long-distance flights into the unknown Antarctic region (Fig. 11).

In addition, serial aerial survey cameras manufactured by Zeiss Aerotopograph (Jena) were built in on both sides for aerial photography flights. These cameras continuously took mostly overlapping pictures of the earth's surface.

Special equipment on the flying boats:

> D-AGAT *Boreas*, built in 1934; structural weight: 6336 kilograms
> D-ALOX *Passat*, built in 1935; structural weight: 6318 kilograms
> 4720-liter fuel tanks for well over 16 flight hours or a range of approximately 2500–2800 kilometers at 150–170 kilometers per hour
> Two serial mapping cameras with film rolls
> 50 metal darts and 10 flags to be dropped out of the planes
> 60 kilograms of water in reserve
> Sea equipment for emergency landings (Fig. 12)[40]

(continued)

---

[40] Schirmacher, Richardheinrich/Mayr, Rudolf: Flüge über der unerforschten Antarktis, in: Ritscher, Alfred (ed.): Wissenschaftliche und fliegerische Ergebnisse der Deutschen Antarktischen Expedition 1938/39. Vol. 1, Leipzig 1942, pp. 232–234.

Apart from the expedition leader (Ritscher), the ship's crew—including Captain Alfred Kottas (Fig. 13), Josef Bludau (physician), and Otto Kraul (ice pilot)—consisted of 59 men in addition to the six crew members needed to operate the catapult and maintain the two airplanes.[41]

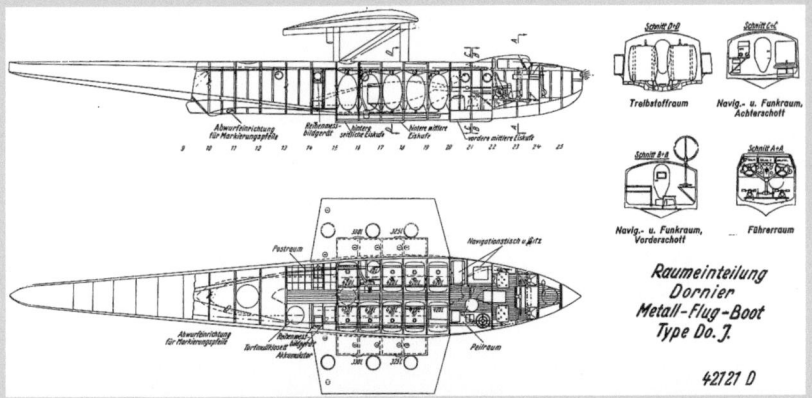

Fig. 11 Spatial division of the Dornier 10-t-Wal flying boats. (Ritscher, Alfred (ed.): Wissenschaftliche und fliegerische Ergebnisse der Deutschen Antarktischen Expedition 1938/39, Leipzig 1942, p. 23)

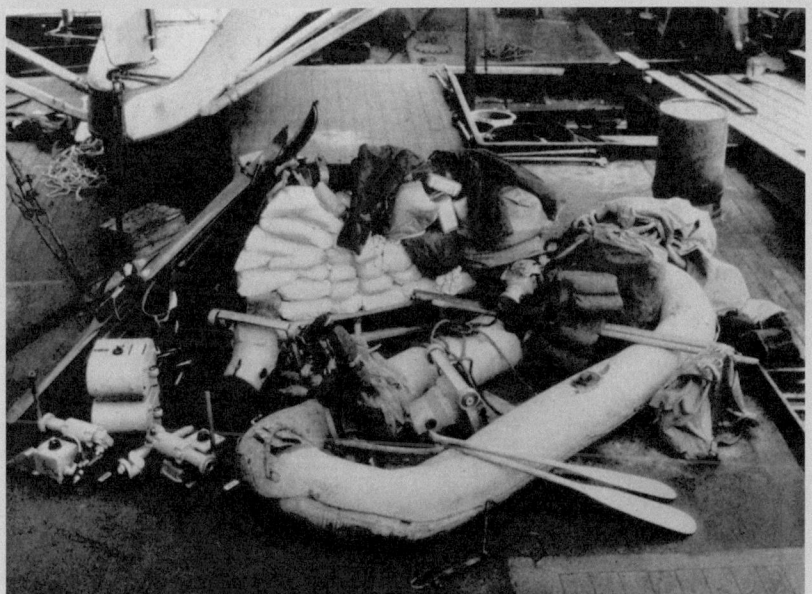

Fig. 12 Additional loading of a flying boat before a long-distance flight. (Leibniz-Institut für Länderkunde, Leipzig, Ritscher estate)

(continued)

---

[41] Ritscher: Wissenschaftliche und fliegerische Ergebnisse, pp. 26–28.

There were also eight men belonging to the flight crew (Figs. 14 and 15), four men for weather services, and four scientists for biological, geophysical, geographic, and oceanographic research; hence, a total of 82 persons were on board the *Schwabenland*.

Fig. 13 Alfred Kottas, the captain of the *Schwabenland*, speaking on the ship's telephone. (Leibniz-Institut für Länderkunde, Leipzig, Ritscher estate)

Fig. 14 Rudolf Mayr, the flight captain of the *Passat*. (Leibniz-Institut für Länderkunde, Leipzig, Ritscher estate)

(continued)

Fig. 15 Richardheinrich Schirmacher, the flight captain of the *Boreas*. (Leibniz-Institut für Länderkunde, Leipzig, Ritscher estate)

The Assignments on the *Schwabenland* Expedition

The organization and outfitting of the third German Antarctic Expedition (1938–1939) differed fundamentally from the two preceding expeditions. It was planned as a brief summer campaign only and had to be successful, come what may.[42] No costs were spared, and the most modern equipment available was provided. The ministries and offices involved were responsible for the scientific staff. In early November, the German War Navy compiled the scientific objectives and their processes in the scope of a Secret Command Matter.[43]

---

[42] Wohlthat, Helmuth: Die Deutsche Antarktische Expedition 1938/39, in: Der Vierjahresplan 3 (1939) 9, p. 614.

[43] Oberkommando der Kriegsmarine, November 3, 1938; Ritscher: Wissenschaftliche und fliegerische Ergebnisse, p. 2, 26.

Throughout the entire voyage, the ship's crew was to survey the topographical relief of the ocean floor by echo sounding. The doctoral candidate Karl-Heinz Paulsen was appointed to carry out the oceanographic surveys. Leo Gburek, another doctoral candidate, who had already taken part in two German Spitsbergen expeditions, was to take over the geomagnetic measurements.[44] The meteorologist Herbert Regula was responsible for studying the weather conditions in high air layers. The Reich Ministry of Food and Agriculture [Reichsministerium für Ernährung und Landwirtschaft] chose the student teacher Erich Barkely to take over whale and bird observations, as well as plankton and krill analyses. Heinz Lange, a probationary teacher, was exempted from work by the Reich Office for Meteorological Services [Reichsamt für Wetterdienst] to take over weather services on board.

The geographer Ernst Herrmann was assigned to the expedition by the Reich education minister (Reichsminister für Volksbildung). In the summer of 1938, Herrmann had conducted an expedition of his own to Spitsbergen, during which an aircraft had been deployed.[45] The ship's officers had to measure the dip of the horizon, test nautical devices, and confirm German nautical maps. Production of coastal panoramas for maritime handbooks was also part of their duties. In the Antarctic itself, the bearing capacity of the ice was to be tested. A cross-section through the South Atlantic Ocean was to be performed by making serial hydrographic measurements up to the islands Trinidade and Martin Vaz [Martim Vaz (in Portuguese)], northwest of Rio de Janeiro. The object of the expedition was to generate a map of the coastal region of Antarctica on the basis of aerial photogrammetric images.

Rudolf Mayr (the pilot of the *Passat*) and his engineer Franz Preuschoff also had some polar experience, as they had already participated in a Danish expedition led by the geologist Lauge Koch in the spring of 1938, in the course of which they had taken photogrammetric images of the Peary Land Peninsula in Northeast Greenland, flying a Dornier Wal.[46] The aerial photographer on the *Passat*, Max Bundermann (a professional photographer for Hansa Luftbild AG), had been part of a Norwegian Svalbard and Arctic Ocean Survey [Norges Svalbard- og Ishavsundersøkelser (NSIU)] expedition to survey East Greenland in 1932, and had taken aerial photogrammetric images of an area of nearly 30,000 square kilometers on only ten flights.[47] His colleague Siegfried Sauer was the youngest man on board and responsible for the aerial photographs taken from the *Boreas*.

As the exploration of the still unknown region in Antarctica was to proceed by aircraft first, special ice darts were to be applied to mark the flyover zones.[48] For this reason, in early November 1938, Ernst Herrmann conducted so-called preliminary trials for drift ice research on the Pasterzen glacier in the Großglockner area (Austria), in which he tested these ice darts, which were to be used in Antarctica to mark the areas

---

[44] Rieche, Herbert: Bericht über die »Deutsche Spitzbergen-Expeditionen 1937 und 1938«, in: Petermanns Geographische Mitteilungen 85 (1939), p. 125.

[45] Herrmann, Ernst: Mit dem Fieseler-Storch ins Nordpolarmeer, Berlin 1942.

[46] Anonymous: Die Insel der Moschus–Ochsen, in: Berliner Nachtausgabe, issue of May 23, 1938.

[47] Barr, Susan: Norway—a consistent polar nation? Oslo 2003, p. 156.

[48] Ritscher: Wissenschaftliche und fliegerische Ergebnisse, p. 7.

explored by the expedition.[49] Afterward, Lufthansa ordered (from Dornier) 500 marking darts with the Nazi emblem engraved on the stabilization surface, which were to be delivered—for reasons of camouflage—for an Arctic expedition.[50] In addition, 20 airdrop flags (flags of the German Reich with a swastika) were also ordered.[51] In the meantime, a second organizational plan had been devised for the expedition and supplemented with a map showing three proposed travel routes (Fig. 16).[52]

The official destination of the expedition, under the auspices of Hermann Göring, was "to secure Germany's right of co-determination and her rightful proportion in the upcoming division of Antarctica among the superpowers by means of an exploratory advance into Antarctic waters and the inland of the Antarctic continent, and thus create the conditions for the Reich's unfettered right of undisturbed execution of whaling essential to the survival of its 80 million people."[53]

Since territorial claims had not been asserted yet for several coastal areas of Antarctica bordering the Atlantic and Pacific Ocean, "the unknown part of Antarctica lying south of Bouvet Island around the 0° meridian was defined as the working area of the expedition" because of its more favorable ice conditions.[54]

One month prior to the departure, Wohlthat determined the final guidelines for the preparation and execution of the German Antarctic Expedition.[55] Accordingly, German Lufthansa was responsible for the sea equipment, the reconstruction of the expedition ship, and the aircraft. In addition, there was a military assignment to explore the islands of Trinidade and Martin Vaz (for use as potential future naval bases). In the scope of the Secret Command Matter, the following questions needed to be answered:

1. Do the islands have anchorages that provide sufficiently good anchoring ground, protection for fuel uptake, equipment of auxiliary cruisers and similar military measures?
2. Are the islands permanently or temporarily inhabited?
3. Are [there] intelligence installations of any kind on the islands?
4. Is [there] a possibility of freshwater supplementation on the islands?

---

[49] Erprobungsstelle der Luftwaffe: October 27, 1938, Erprobungsstelle der Luftwaffe, Programm für Auftrag E728/11, Leibniz-Institut für Länderkunde, Leipzig, Ritscher estate, File AuE1, Abt. E; Ritscher: Wissenschaftliche und fliegerische Ergebnisse, pp. 6–7.

[50] Deutsche Lufthansa: November 12, 1938, Brief an Dornier, Leibniz-Institut für Länderkunde, Leipzig, Ritscher estate, File AuE1, Abt. D.

[51] Ritscher, Alfred: December 1, 1938, Leibniz-Institut für Länderkunde, Leipzig, Ritscher estate, File AuE1, Abt. B.

[52] Wohlthat, Helmuth: November 10, 1938, W XVII/163, Leibniz-Institut für Länderkunde, Leipzig, Ritscher estate, File Bh1, Abt. VJP; Oberkommando der Kriegsmarine: November 11, 1938, B.No. 1087/37 g.Kds. BH W V, III. Ang., Leibniz-Institut für Länderkunde, Leipzig, Ritscher estate, File Bh1, Abt. OKM.

[53] Ritscher: Wissenschaftliche und fliegerische Ergebnisse, p. 2.

[54] Wohlthat: Die Deutsche Antarktische Expedition, p. 614.

[55] Wohlthat, Helmuth: November 21, 1938, Brief an Ritscher und andere beteiligte Stellen, W XVII/175, Leibniz-Institut für Länderkunde, Leipzig, Ritscher estate, File Bh1, Abt. VJP.

Fig. 16 The proposed routes of the third German Antarctic Expedition (1938–1939) in the Atlantic along the meridian of 0° and in the Pacific along the meridians of 90°W or 120°W. (Leibniz-Institut für Länderkunde, Leipzig, Ritscher estate)

5. Do the islands have fauna and flora that could be used to supplement food rations, for example, for submarines?
6. Are the islands occasionally approached by any kinds of vehicles? If yes, for what purpose?[56]

---

[56] Oberkommando der Kriegsmarine: November 21, 1938, B.No. 2215/38 g. Kds. BH W V; Leibniz-Institut für Länderkunde, Leipzig, Ritscher estate, File Bh1, Abt. OKM.

A direct military interest in Antarctica itself cannot be gleaned from the documents in the archives—contrary to the later creation of numerous myths.

Since the Kaiser Wilhelm Society had chosen not to be the official support organization for the expedition, Wohlthat appointed the German Research Foundation [Deutsche Forschungsgemeinschaft] as the support organization ten days before the departure and took the necessary money from the funding for the Four-Year Plan.[57] Because any preparations for an intended occupation in Antarctica had to be kept secret, the expedition was to display a purely scientific character on the outside.[58] Apart from the general weekly reports sent to German Lufthansa, however, all other messages were to be transmitted in secret code and then forwarded by Wohlthat after being deciphered.[59] To this end, an Enigma encryption machine had been taken on board the ship.

The Execution of the German Antarctic Expedition (1938–1939)

On December 17, 1938, the expedition set out from Hamburg without a big farewell.[60] Radiosondes were launched on a regular basis as early as on the second day in order to measure air pressure, temperature, and humidity (with the instruments attached to the balloon) and transmit the data by radio to a recipient on board the *Schwabenland*. In this way, a good picture of the meteorological conditions in the stratosphere was obtained up to an altitude of 30 kilometers (Figs. 17 and 18).

When the expedition passed Spain's Cape Finisterre on December 21, 1938, nothing could be seen of the fighting in the still-ongoing Spanish Civil War in which the National Socialist Condor Legion had taken part by destroying Guernica in 1937. Instead, the routine of a research expedition prevailed on board. Shortly before Christmas, the messrooms were decorated for the great festive meal. In the common room stood two fresh little Christmas trees, festooned with lights, and two tables with numbered gifts to be raffled in the course of the evening. Reich flags with swastikas, the house flag of the expedition, and colorful signal flags adorned the walls. Three good bottles of beer and a bag containing nuts and fruit for everyone were on the tables. The party was "both solemn and jolly at the same time … and the Christmas speech of Reich Minister Heß [reached the crew] here below the Canary Islands, too."[61] However, the atmospheric disturbances were so strong that

---

[57] Ritscher: Wissenschaftliche und fliegerische Ergebnisse, p. 14; Wohlthat, Helmuth: December 8, 1938, W XVII/200, R73 No. 242, Bl. 2, Bundesarchiv, Koblenz.

[58] Deutsche Forschungsgemeinschaft: December 8, 1938, Deutsche Forschungsgemeinschaft, Vermerk, R73 No. 242, Bl. 7–8, Bundesarchiv, Koblenz.

[59] Wohlthat: December 10, 1938.

[60] Herrmann: Deutsche Forscher, pp. 16–27; Ritscher: Wissenschaftliche und fliegerische Ergebnisse, pp. 28–37.

[61] Herrmann: Deutsche Forscher, p. 27.

Fig. 17 Ascent of a radiosonde. (*Stuttgarter Illustrierte* Nr. 37 issue from 10 September 1939)

the broadcast had to be suspended, to the dismay of the political leader of the *Schwabenland*, Second Officer Karl-Heinz Röbke, who was responsible for ideological education and surveillance of discipline in the sense of the Nazi regime. The expedition leader (Ritscher) then entertained the audience with a report about the rescue of the Schröder-Stranz expedition in Spitsbergen, which had taken place 26 years previously. The band on board provided the background for the party, which ended at 2:00 a.m. (Fig. 19).

In contrast to other expeditions in the Nazi period—for example, the simultaneous Tibet expedition (1938–1939) of Ernst Schäfer in search of the alleged origin of the Aryans—the Nazi ideology did not seem to be of great importance to the research of the *Schwabenland* expedition.

On the day after Christmas Day, the *Schwabenland* reached the African coast.[62] The line-crossing ceremony on New Year's Eve was exploited for a great deal of ceremonial nonsense. The 26 people to be baptized received cookies baked with turpentine and raspberry juice flavored with petroleum (Fig. 20). The medical

---

[62] Ibidem, pp. 31–35; Ritscher: Wissenschaftliche und fliegerische Ergebnisse, pp. 37–45.

Fig. 18 Evaluation of a radiosonde ascent in the tropics. (*Die Weite Welt* Nr. 28 issue from 9 July 1939)

doctor then tested each man's hearing, coating the men's ears with fresh oil paint in the process. Then the men's beards, lathered with lubricating grease, were shaved. At last, jet-black figures dunked the candidates three times under water, after which they too were black from head to toe.

When the expedition reached the British volcanic island of Ascension in the South Atlantic on January 2, 1939, regular echo soundings of the ocean depth commenced (Fig. 21). From Tristan da Cunha, which was also a possession of the British Empire, the expedition took a southeasterly course to Bouvet Island, which belonged to Norway. Now the scientists discussed the work program and the organization of

Fig. 19 Christmas celebration in the common room on December 24, 1938. The expedition leader, Alfred Ritscher, is *standing in the middle*. (Leibniz-Institut für Länderkunde, Leipzig, Ritscher estate)

Fig. 20 The line-crossing ceremony on New Year's Eve in 1938. (Deutsches Schifffahrtsmuseum, Bremerhaven)

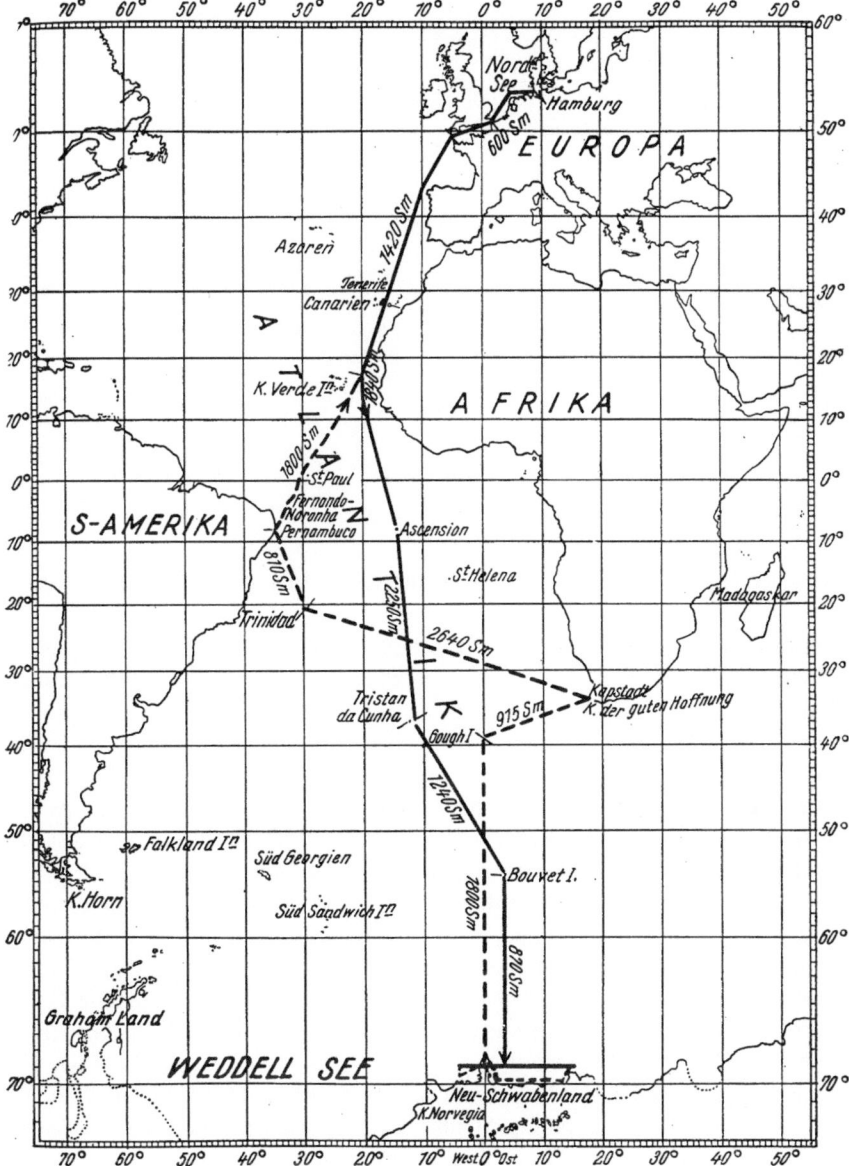

Fig. 21 The route map of the third German Antarctic Expedition on the *Schwabenland*. (Ritscher, Alfred (ed.): Wissenschaftliche und fliegerische Ergebnisse der Deutschen Antarktischen Expedition 1938/39, Leipzig 1942, p. 29)

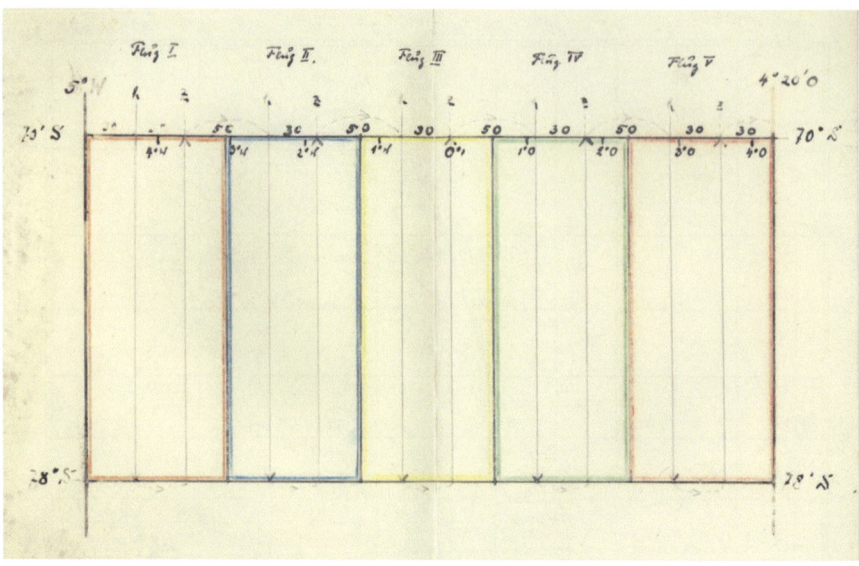

Fig. 22 The plan of the flight route. (Leibniz-Institut für Länderkunde, Leipzig, Ritscher estate)

flights with the aircraft crew and the catapult leader.[63] The unknown region of Antarctica between 70°S and 78°S and between 5°W and 4°20′E was to be measured photogrammetrically, optimally by applying a quadrangular flight pattern—in other words, it was to be photographed by using serial mapping cameras. The calculation of the required film material was based on a flight strip 880 kilometers in length (a flight southward), 30 kilometers in width (a right-angled flight eastward), and another 880 kilometers in length (a right-angled flight northward) (Fig. 22). The next flight was to begin 50 kilometers farther east in order to capture as much ground surface as possible with a certain degree of overlap. Only a small strip below the aircraft could not be photographed. At flight altitudes of 3000 or 1000 meters, areas of 493,000 or 176,000 square kilometers, respectively, could be documented (Fig. 23).

Although there should have been much more ice south of the Antarctic Circle, according to earlier reports, only one lone iceberg was adrift in the far distance (Figs. 24 and 25). Soon they saw the first whale and finally, on January 19, 1939, they reached the pack ice lying in front of the ice shelf border at 69°14′S and 4°30′W.[64] As the ship's hull had been outfitted with simple ice strengthening only, the *Schwabenland* had to stay, if possible, in open waters at all times.

---

[63] Ritscher, Alfred: (January 1939), Organisationsplan für die beabsichtigten Antarktisflüge. Leibniz-Institut für Länderkunde, Leipzig, Ritscher estate, File Bb1, Abt. Flüge.

[64] Kraul, Otto: Käpt'n Kraul erzählt. 20 Jahre Walfänger unter argentinischer, russischer und deutscher Flagge in der Arktis und Antarktis, Berlin 1939, pp. 229–230; Regula, Herbert: Die Wetterverhältnisse während der Expedition und die Ergebnisse der meteorologischen Messungen, in: Ritscher, Alfred (ed.): Deutsche Antarktische Expedition 1938/39. Wissenschaftliche Ergebnisse Vol. 2, Issue 1, Hamburg 1954, p. 28.

Fig. 23 The two aerial photographers Max Bundermann and Siegfried Sauter in the tropics. (Leibniz-Institut für Länderkunde, Leipzig, Ritscher estate)

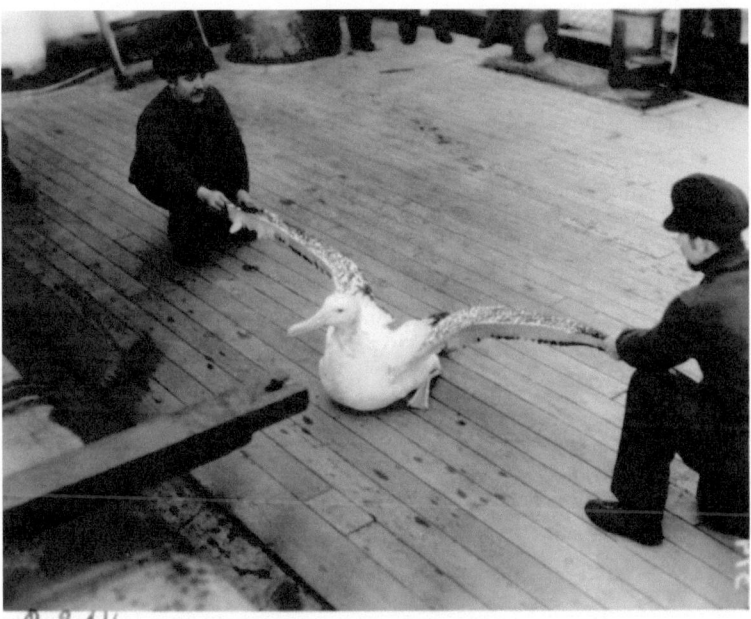

Fig. 24 Measuring an albatross on board the *Schwabenland*. (Leibniz-Institut für Länderkunde, Leipzig, Ritscher estate)

Fig. 25 A table iceberg. The *Schwabenland* can be seen *in the foreground*. (Leibniz-Institut für Länderkunde, Leipzig, Ritscher estate)

However, the expedition had arrived in the south too late, since the area between Coats Land in the west and 45°E, including the German working area, had already been placed under Norwegian sovereignty by a royal decree.[65] How could this have happened? The starting point was a trip by Adolf Hoel (the director of the NSIU) to Europe in December 1938, which had also led him to Berlin to make preparations for a polar exhibition planned to take place in Bergen in 1940.[66] Hoel's attention had already been drawn to Germany's interest in occupying territory in Antarctica in early December 1938, owing to a visit from Dr. Lehmann, an attaché at the German Embassy in Oslo. Lehmann asked him questions pertaining to the sectoral principle (which, in the Arctic, describes the extension of state borderlines on continental coasts up to the North Pole) and about land occupations in Antarctica.[67] Lehmann underscored the private nature of his question, saying he was personally interested in studying international law. When Hoel invited Lehmann a second time, he suddenly mentioned to him that only regions between 20°E and 40°E and/or between 80°W and 130°W would come into question in the event of a German occupation in Antarctica (see Fig.16). Hoel emphasized that both regions would lie in Norway's

---

[65] Wohlthat, Helmuth: January 17, 1939, Telegramm an Ritscher auf der Schwabenland, Leibniz-Institut für Länderkunde, Leipzig, Ritscher estate, File AA1, Abt. Norwegen.

[66] Lehmann: December 10, 1938, Aktennotiz, Leibniz-Institut für Länderkunde, Leipzig, Ritscher estate, File AA; Barr: Norway, pp. 170, 194; Abs, Otto: Professor Adolf Hoel 80 Jahre alt, in: Polarforschung 27 (1957), p. 52.

[67] Lehmann: December 10, 1938.

sphere of interest, even though Norway had not yet put forward any claim. As Hoel did not speak German very well, Lehmann did not exactly know whether Hoel had gained intelligence about Germany's occupation plans or whether he was simply trying to answer his questions.

Lehmann's report about his meeting with Hoel was forwarded to Wohlthat, who dispatched it to Ritscher along with a cover letter dated December 14, which referred only to the planned American Antarctic expedition led by Lincoln Ellsworth—another subject Lehmann had discussed at length in his report. Lehmann made no mention to the effect that Hoel might have had any suspicions.

When Hoel traveled to Berlin in December, he tried to contact Ernst Herrmann, whom he knew from his expedition to Spitsbergen. Herrmann's wife told Hoel on the telephone that her husband had been on a polar expedition since December 17.[68] Since Arctic expeditions do not take place in the winter, it stood to reason that Herrmann was on his way to Antarctica during the southern summer—in particular, to the region south of Cape Town, which Norway also wanted to take possession of on the basis of several years of coastal research. Thus, after his return on December 22, Hoel alerted the Norwegian Ministry of Foreign Affairs, with the result that a huge conference, involving many experts on Antarctica, was held on January 3. As early as January 14, everything was prepared, and Norway officially announced that it was taking possession of Dronning Maud Land, although this region was not permanently inhabited and nor had its inland been explored, as would be required to take possession of land in accordance with international law.

Despite the fact that the Norwegian Ministry of Justice had been authorized to exercise police powers within her possession in Antarctica, Wohlthat instructed Ritscher to carry out the commissioned expedition assignment unaltered.[69] Hence, the expedition continued with its program and its aircraft were made ready to take off on their first reconnaissance fight, primarily to test the photographic devices and explore the sea ice.[70] Once the *Boreas* was in the air, the radio operator sent the first information in Morse code relating to the distribution of ice. He had seen an ice-free wake further westward, which would be best suited to a safe landing, and to which the *Schwabenland* now set her course. After one hour, the *Boreas* returned. In the meantime, preparations had been made for the first long-distance flight, which was to take off on January 20 at 4:40 a.m. in bright sunlight.

Since the risk associated with landing on the mainland was too high, astronomical position findings on the earth's surface had not been planned. As a consequence, only the position of the *Schwabenland* could be drawn upon as a reference point, along with in-flight recorded data pertaining to the flight direction and velocity, in order to allocate the photographs retrospectively to geographical coordinates.

---

[68] Barr: Norway, p. 170; Orheim, Olav: How Norway got Dronning Maud Land in: Winther, Jan-Gunnar (ed.): Norway in the Antarctic—from Conquest to Modern Science, Oslo 2008, pp. 44–59.
[69] Vierjahresplan: January 17, 1939, Notiz, Leibniz-Institut für Länderkunde, Leipzig, Ritscher estate, File Bh1, Abt. VJP.
[70] Ritscher: Wissenschaftliche und fliegerische Ergebnisse, p. 48.

For all flights, good weather conditions were of fundamental importance.[71] To forecast the weather, observations of air pressure, temperature, humidity, wind direction, wind speed, and precipitation were carried out every six hours, using the onboard weather station, starting at midnight.[72] These so-called term data were transmitted by radio to the overseas radio broadcasting station in Quickborn, 30 kilometers north of Hamburg, where messages were also received from whaling and merchant ships operating in the South Atlantic. These synoptic data were gathered and sent back to the *Schwabenland* every evening at 8:00 p.m. In addition, the meteorologists released a radiosonde, which produced additional flight altitude weather data at least once daily.[73]

Along with weather reports from South America, the meteorologist compiled this information daily to create maps for aviation weather forecasts on-site. In addition, the turning of the wind direction with height was determined with pilot balloons for current flight weather briefings.

Prior to their first long-distance flight on the morning of January 20, the aircraft crew was briefed at 3:00 a.m. with the current weather map and received information about the general weather situation, including a forecast of the wind speed and direction, the expected cloud types and their altitude, and perhaps also any potential precipitation, because of the risk of aircraft icing.[74]

The forecast for the Antarctic mainland revealed that the good weather conditions would persist for the next 15 hours, although the available daylight needed for taking photographs would probably provide sufficient brightness for only ten hours. Finally, at 4:38 a.m., the *Boreas* was launched by the catapult on her first trip into the unknown. Afterward, the *Passat* was hoisted onto the catapult and prepared for a launch in case the *Boreas* became involved in an emergency situation—a procedure that took four hours.[75] In the meantime, Ritscher and the other scientists waited with excitement for the first radio messages about the potential discoveries beyond the southern horizon. Ernst Herrmann had a blank sheet of paper, with coordinates laid out before him, on which he would enter the data from the *Boreas*, in order to be able to locate the aircraft more easily in the event of an emergency or a crash landing (Fig. 26).

---

[71] Regula: Die Wetterverhältnisse, pp. 25–27.

[72] Regula, Herbert: Die Arbeiten der Expeditionswetterwarte. Part I: Terminbeobachtungen, Höhenwindmessungen, Wetterdienst, Sonderuntersuchungen, in: Vorbericht über die Deutsche Antarktische Expedition 1938/39, Annalen der Hydrographie und Maritimen Meteorologie, VIII. Sup-ple-ment 1939, pp. 33–35.

[73] Lange, Heinz: Die Arbeiten der Expeditionswetterwarte. Part II: Radiosondenaufstiege, in: Vorbericht über die Deutsche Antarktische Expedition 1938/39, in: Annalen der Hydrographie und Maritimen Meteorologie VIII. Supplement 1939, pp. 35–36.

[74] Herrmann: Deutsche Forscher, p. 59; Ritscher: Wissenschaftliche und fliegerische Ergebnisse, pp. 50–85; Regula, Herbert: February 2, 1939, Wetterberatung für einen Flug von der »Schwabenland« in südöstlicher Richtung, Leibniz-Institut für Länderkunde, Leipzig, Ritscher estate, File Bb1 Meteorologie.

[75] Ritscher: Wissenschaftliche und fliegerische Ergebnisse, pp. 50–53; Schirmacher/Mayr: Flüge, pp. 249–250; Herrmann: Deutsche Forscher, pp. 62–65.

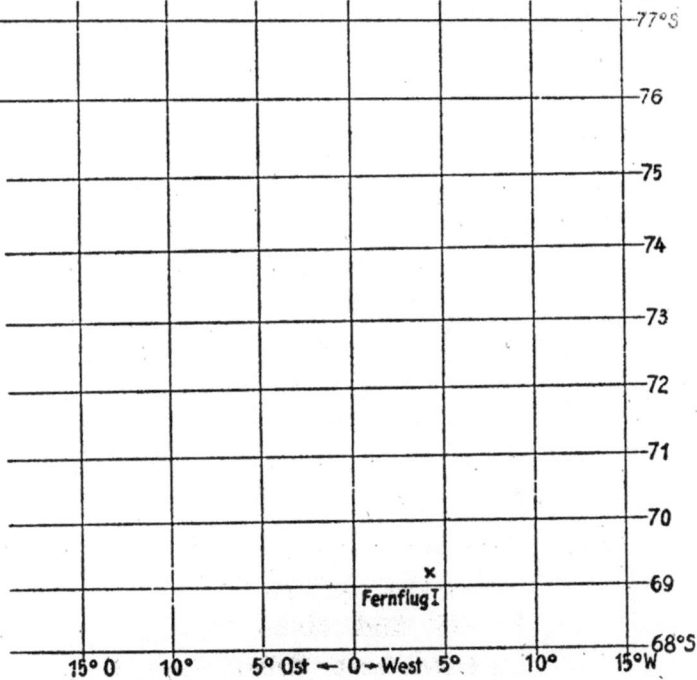

Fig. 26 Antarctica—a map still undrawn. The *X* marks the launching site of the first long-distance flight into the unknown. (Herrmann, Ernst: Deutsche Forscher im Südpolarmeer, Berlin 1941, p. 61)

After 29 minutes, the *Boreas* flew over the low ice shelf boundary at an altitude of 1700 meters. Shortly afterward, the crew discovered the first mountains in the southeast. Otherwise, everything lay under a white cover of snow, which, in some places, displayed drifts. The first names were assigned at 6:30 a.m., when the first ice-free elevations (nunataks)—Kugel [meaning "ball" or "bullet"] and Kegel [meaning "cone"]—lay abeam. Behind them, more mountains came into view (Fig. 27). The board mechanic Franz Preuschoff captured the discoveries in drawings while the flight was in progress, and Herrmann marked these elevations on his working map.

One hour later, the flight passed the extreme western boundary of the mountain, which extended in a south–southwestern direction. The ice cover increased further during the flight.[76] At about 8:00 a.m., the crew was confronted by a mountain range. At an altitude of about 2500 meters, the flight passed a steep rocky ridge on the starboard side (Fig. 28). In the fold behind it was fog. The inland ice that followed increased more and more, and soon reached an estimated altitude of more than 4000 meters. As the crew's sight was limited by rising fog shortly after they passed the mountain range at an altitude of only 100 meters above the ground,

---

[76] See also the detailed description of the flight in Mayr, Rudolf: Deutsche Flugboote fliegen über die Antarktis, in: Der Lufthanseat 3 (1939) 6, pp. 1–6.

Fig. 27 The Lutz Ridge in Ritscher Land, photographed in 1939. (Leibniz-Institut für Länderkunde, Leipzig, Herrmann estate)

Schirmacher aborted the flight at 73°52'S and 4°45'W (approximately 600 kilometers south of the *Schwabenland*) in order to conduct the return flight 30 kilometers further east as planned.

In this situation, even a small saving in weight was essential for the safety of the flight crew in order for them to quickly rise again to a safe altitude. Given the mountain range in front of them, the crew jettisoned all expendable items of ballast at once, among which were all of the marking darts and flags.[77] The planned site marking of the turning point could therefore not be done. Schirmacher's flight report, which he wrote afterward, therefore hardly mentioned any details or exact locations: "Darts and flags were dropped as ordered."[78] It never became known that this

---

[77] Sauter, Siegfried: May 25, 1992, Brief an Cornelia Lüdecke, private possession Lüdecke, Munich.

[78] Schirmacher, Richardheinrich: January 20, 1939, Flugprotokoll, Leibniz-Institut für Länderkunde, Leipzig, Ritscher estate, File Bb1, Abt. Flüge; identical reprint of the report in Schirmacher/Mayr: Flüge, p. 250.

The Execution of the German Antarctic Expedition (1938–1939) 167

Fig. 28 Mayr Kette (now known as Jutulsessen) and the Mühlig-Hofmann Mountains. (Leibniz-Institut für Länderkunde, Leipzig, Herrmann estate)

had been done only once throughout the entire expedition.[79] No other flight report mentioned any objects being dropped from the air, although it might be assumed with certainty that the coordinates of the dropped darts, or at least the distinctive turning points marked with the flags, would have been indicated in the flyover drawings and in the flight report.[80] In Herrmann's travel report, however, (fictitious) markings displaying swastika flags were drawn on Map 2 at the end of the book and Ritscher dutifully mentioned the aerial droppings in his travel report.[81] In his original drawing, which gave an overview of all flights, he marked only three locations with a cross in a circle, indicating where swastika flags had been stuck into the snow during landings (Fig. 29). Appearances had to be upheld, at any rate; thus, officially, everyone remained silent on this matter.

---

[79] Conversation of Cornelia Lüdecke with Ilse Ritscher, who was secretary of the expedition office at that time, during the International Polar Conference in Gottingen on April 12, 1991. According to Mrs. Ritscher's statement, Schirmacher's procedure after his return had stirred up tremendous anger on board, as presumably there was no further dropping material available for the next flights.

[80] See also Schirmacher/Mayr: Flüge, und Originalberichte im Leibniz-Institut für Länderkunde, Leipzig, Ritscher estate Akte Bb1, Abt. Flüge.

[81] Herrmann: Deutsche Forscher; Ritscher: Wissenschaftliche und fliegerische Ergebnisse, pp. 80–81.

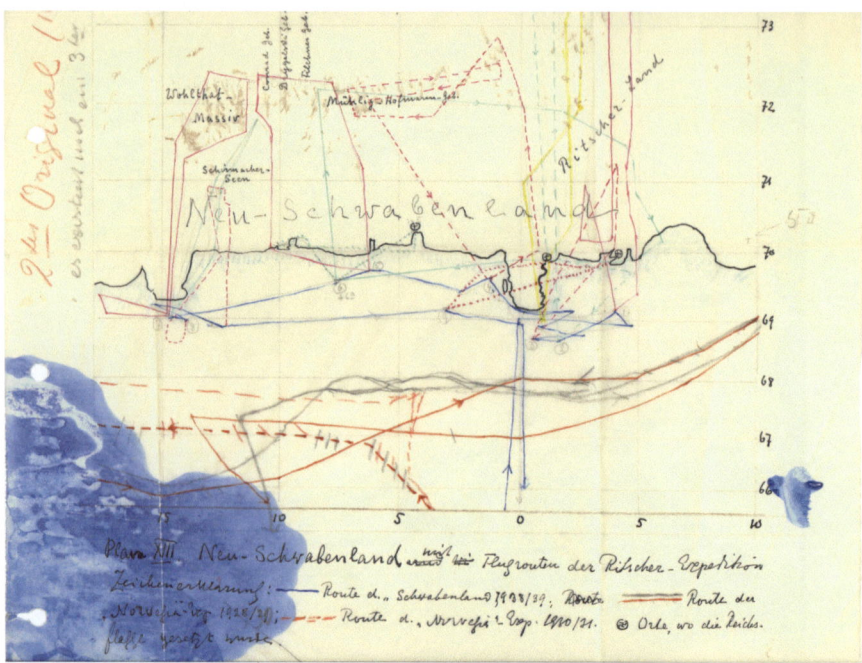

Fig. 29 Ritscher's overview of all flights over Neu-Schwabenland and the navigation routes. (Leibniz-Institut für Länderkunde, Leipzig, Ritscher estate)

Since the weather remained fair on January 20, a subsequent second photographic flight of the *Passat* was agreed to by radio communication at 11:51 a.m., leading from 2°40'W southward to the eastern border of the mountain.[82] However, the starboard camera broke down after 40 min, so the aircraft had to turn around again. On its return flight, the aircraft flew only 100 meters above the ground in order to study the surface of the inland ice in more detail. In many crevasses, it gradually dropped northward into a kind of fjord (Fig. 30), with an ice-free wake of 1200 meters by 200 meters, in which seals and penguins were sighted. This wake was excellently suited to a landing. Upon returning to the *Schwabenland*, the *Passat* was to search for a navigable way out of the ice, which was floating all around by now. This could be easily accomplished from the air (Fig. 31).

While waiting, the oceanographer measured the temperature and salt content at various ocean depths and hauled up a ground sample from the ocean floor 2000 meters below. The biologist caught plankton in a net in order to study the starting point of the food chain.

After the aircraft returned, they were maintained by the mechanics for the next flights (Figs. 32 and 33). The mechanics also repaired the malfunctioning serial mapping camera. The new launching position was farther in the east. The weather was still good, but messages from whaling ships in the surrounding area indicated

---

[82] Ritscher: Wissenschaftliche und fliegerische Ergebnisse, pp. 52–96; Schirmacher/Mayr: Flüge, pp. 249–265; Herrmann: Deutsche Forscher, pp. 71–89.

The Execution of the German Antarctic Expedition (1938–1939)

Fig. 30 Glacier crevasses. (Leibniz-Institut für Länderkunde, Leipzig, Herrmann estate)

Fig. 31 The hitched flying boat *Passat* is hoisted on board the *Schwabenland* in 1939. (Leibniz-Institut für Länderkunde, Leipzig, Ritscher estate)

Fig. 32 Ritscher on the catapult deck of the *Schwabenland*. (Leibniz-Institut für Länderkunde, Leipzig, Ritscher estate)

that there would be a change in the weather within the next few days. The *Passat* and the *Boreas* now flew alternately and discovered more and more mountain ranges. On the second long-distance flight, on January 21, the Matterhorn presented a very characteristic landmark (Fig. 34).

As the flight procedures gradually became routine, the second plane could take off on a reconnaissance flight, with one guest in the photographer's seat, when the first was on its way back and only one or two flight hours away from the ship. The first "joy ride" lasted almost 3½ hours because the expedition leader wanted to see the nearby nunataks Kugel and Kegel with his own eyes. At the same time, Barkley, Gburek, Herrmann, and the electrical engineer Herbert Bruns took a boat to an ice floe to carry out ice analyses. Gburek wanted to take magnetic measurements, but his attempts were futile because the floe was rocking too much as a result of the swell.

After strenuous days during which most of the crew had worked from 3:00 a.m. or 4:00 a.m. till midnight, all were glad to see the first spell of bad weather, which meant they could now finally recover for a week (Fig. 35).

On January 29, 1939, Schirmacher took off again on the *Boreas*. On the fourth long-distance flight, the crew discovered new mountain ranges and special mountain forms, which they named Hasenrücken [Rabbit's Back], Hohenstaufen, and Kubus. Schirmacher estimated that the height of the inland ice in the south was about 4000 meters. This time, a second special flight of the *Passat*, with Herrmann as the guest on board, was scheduled quite early in order to test the first external landing on the edge of the ice shelf (Fig. 36).

Fig. 33 The *Boreas* on the hook. (Leibniz-Institut für Länderkunde, Leipzig, Ritscher estate)

Fig. 34 A drawing of the Matterhorn. (Leibniz-Institut für Länderkunde, Leipzig, Ritscher estate)

Fig. 35 Ritscher (*second from left*) talking to expedition members, with Kraul *on the far right*. (Leibniz-Institut für Länderkunde, Leipzig, Ritscher estate)

Fig. 36 Geographer Ernst Herrmann, photographed around 1937. (Leibniz-Institut für Länderkunde, Leipzig, Herrmann estate)

The Execution of the German Antarctic Expedition (1938–1939)   173

The fjord was nearly two kilometers long and, at the landing site, 500 meters wide. Mayr determined the position of the landing site right after the *Passat* had moored at the edge of the ice (Figs. 37 and 38). Herrmann took a group picture of the aircraft crew and measured the ocean depth by echo sounding.

On their return flight, the crew took three Adélie penguins and an emperor penguin on board for Tierpark Hagenbeck in Hamburg, as well as a seal they had killed. A basin and a shed were carpentered to house the penguins on the deck of the *Schwabenland*, and Barkley assumed responsibility for their further care.

By the time of the fifth photography flight on January 30, the flights had already become a routine operation (Figs. 39 and 40). The outside temperature at a maximum flight attitude of 3850 meters was −19 °C when the aircraft flew in a full circle around the Drygalski Mountains. On the next special flight of the *Boreas*, an ice floe suitable for magnetic measurements was found. In addition, an ice hill at 70°17'S and 41°22'E was marked with a swastika flag when the aircraft landed at the ice boundary.

The sixth photography flight on January 31 had to be discontinued after 4½ hours because of a sudden deterioration in the weather (Fig. 41). On its way back, the *Passat* flew a reconnaissance mission over a broad coastal area and made an external landing, during which five emperor penguins were captured and loaded, against their will, into the Dornier Wal.

The seventh photography flight could not take place before February 3. As had previously been the case, the trimming of the aircraft regularly broke down because of the low temperature of −31 °C that prevailed at an altitude of over 4000 meters, but this time even more instruments failed. However, since the visibility was good and the terrain was flat enough for emergency landings, Mayr decided to continue flying despite the technical failures, although he would never have done so under ordinary circumstances. This time, they flew around the Wohlthat Massif [Wohlthat Mountains], which lay in the eastern part of the working area and had the highest mountain summits in this region (Fig. 42). Only on the return flight, when the temperature was −7 °C at an altitude of 1000 meters, did all instruments function normally again. The *Passat* was subsequently decommissioned because of these defects.

During a reconnaissance flight on the same afternoon, Ritscher wanted to look at the eastern side of the overflown area, and this produced the greatest surprise of the expedition, for on the return flight they became aware of a small mountain range that "apparently displayed free water sites."[83] Upon approaching the area, it turned out to be a glacier moraine. In deep cracks, water flowed through the ice, where, according to previous opinion, everything should be frozen. In addition, there were small ponds in the depressions between the rocks. The rock around them was free of snow and ice. This Boreasic Lake District [Boreasische Seenplatte]—or Schirmacher Oasis [Schirmacher-Oase], as it was later called—was at 70°45'S and 11°40'E, exactly on the boundary between the inland ice and the ice shelf, and at about 150 meters above sea level (Fig. 43).

---

[83] Schirmacher/Mayr: Flüge, p. 263.

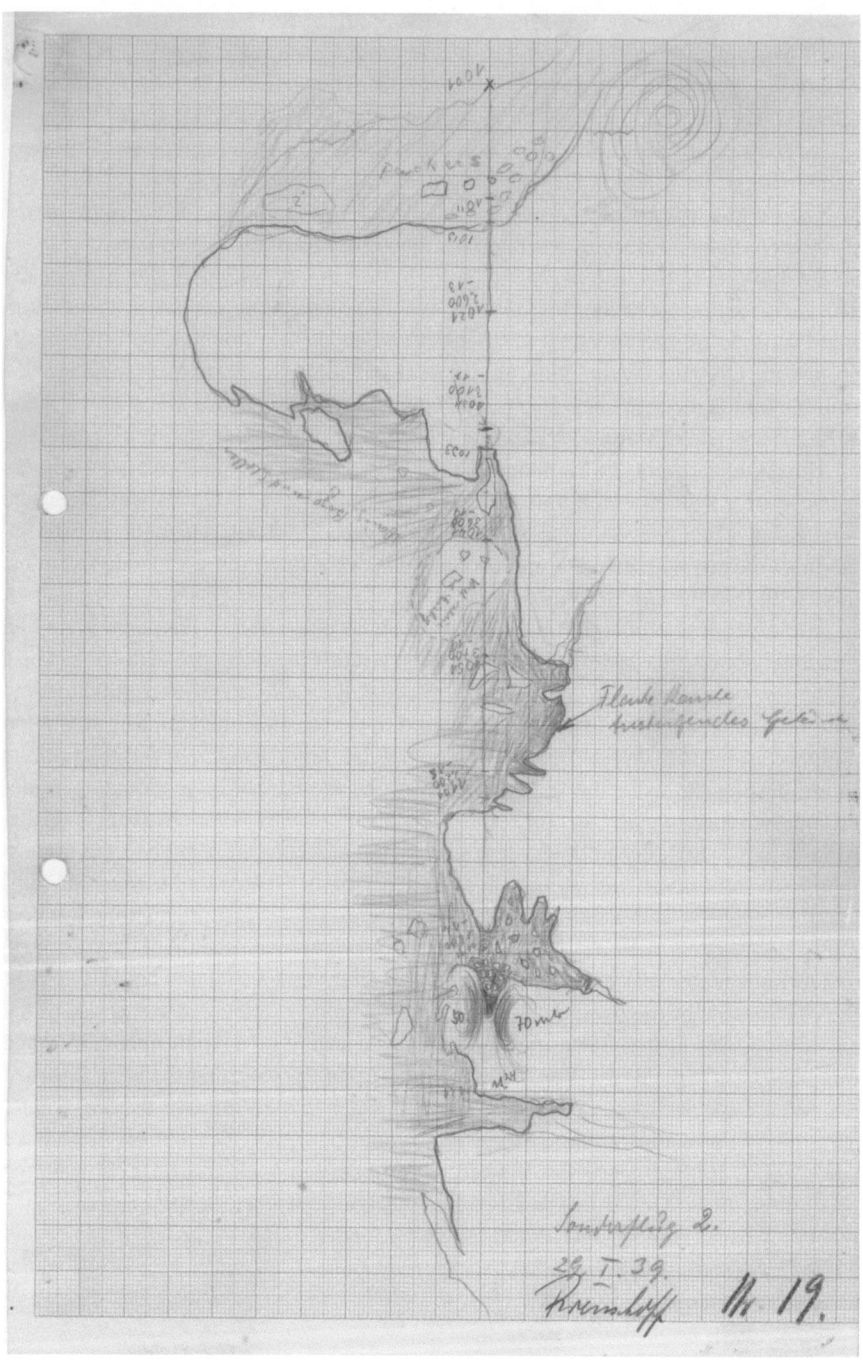

Fig. 37 A drawing from the second special flight along the coast. (Leibniz-Institut für Länderkunde, Leipzig, Ritscher estate)

Fig. 38 The *Boreas*, moored at the ice boundary. (Leibniz-Institut für Länderkunde, Leipzig, Ritscher estate)

At an outside temperature of −5 °C, the open water appeared all the more strange, since the inland ice is made of freshwater and not salty seawater, which freezes at −1.9 °C.[84] Unfortunately, the lakes were not large enough for the flying boat to land on them and the nearby ice surface was too rough for a landing, so a closer inspection could not be done. An aircraft designed for ice landings could establish the starting point for later land expeditions at this site. After the return of the flight, this sensational discovery was discussed with great excitement. Were these lakes perhaps a product of insolation?[85] Herrmann, who had already occupied himself with volcanoes in Iceland, interpreted the lakes as a product of volcanism.[86] A more exact evaluation of the aerial images later showed that the lakes had formed in basins in which meltwater from sun-exposed northern slopes had been dammed.[87]

After the second aircraft had broken down, no further long-distance flights could be scheduled, for safety reasons. However, this did not matter, since the "objective of the expedition was entirely fulfilled in the east, and almost entirely in the west."[88]

---

[84] Ritscher: Wissenschaftliche und fliegerische Ergebnisse, p. 75.
[85] Schirmacher/Mayr: Flüge, pp. 244–245.
[86] Herrmann: Deutsche Forscher, pp. 88–89.
[87] Ritscher, Alfred: Oasen in Antarktika, in: Polarforschung 16 (1946) 1/2, pp. 70–71.
[88] Ritscher: Wissenschaftliche und fliegerische Ergebnisse, p. 76.

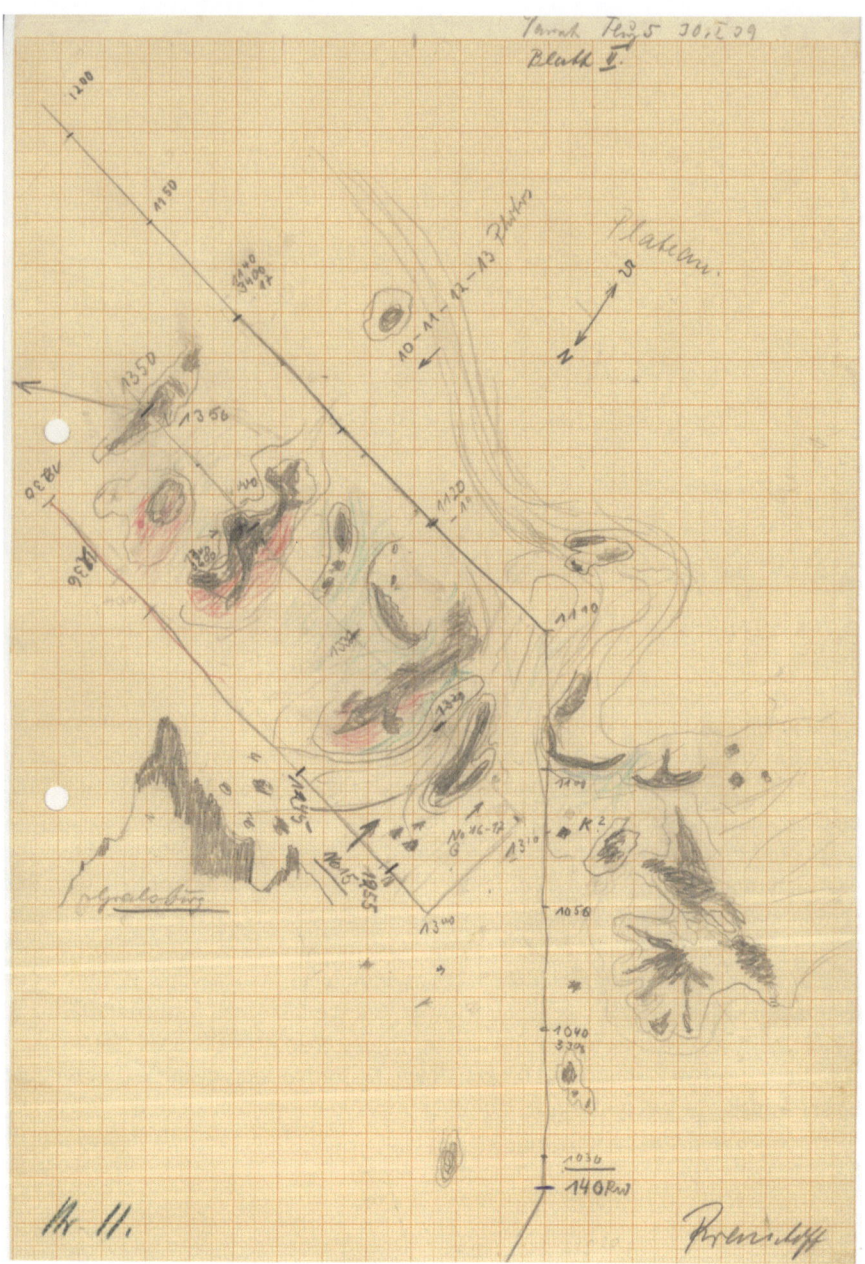

Fig. 39 The fifth long-distance flight, on January 30, 1939, between 10:30 and 13:50. The drawing includes the position of Gralsburg between 12:45 and 12:55. (Leibniz-Institut für Länderkunde, Leipzig, Ritscher estate)

Fig. 40 The so-called Gralsburg as seen from the northwest. The eastern flank of the Filchner Mountains extends *from the center of the picture to the right*. The Drygalski Mountains and the Matterhorn are visible *in the background*. (Leibniz-Institut für Länderkunde, Leipzig, Herrmann estate)

For this reason, an eighth photography flight of a shorter duration was scheduled to take place in order to photograph the lake district.[89]

The season was closing. The very last flight of the expedition took place on February 5. This time, the *Boreas* was flown by the rested crew of the *Passat*, who additionally had taken the meteorologist Lange and the oceanographer Paulsen on board. After they had fastened the flying boat to the ice shelf, using an ice anchor, they caught two penguins, among which was a small gray specimen, which, on the *Schwabenland*, was initially considered to be a new species and was named Agathe (after the radio call sign of the D-AGAT *Boreas*). However, it soon turned out that it too was an emperor penguin, still in molt. With this catch, the expedition came to an end because the weather deteriorated more and more and the pack ice near the ice shelf coast was getting denser.

In his fourth weekly report, Ritscher informed the commissioner of the Four-Year Plan about the results of the expedition.

---

[89] Schirmacher/Mayr: Flüge, pp. 263–265; Herrmann: Deutsche Forscher, p. 82; Ritscher: Wissenschaftliche und fliegerische Ergebnisse, pp. 78–80.

Fig. 41 The northern foothills of the Conrad Mountains, viewed from the north. (Leibniz-Institut für Länderkunde, Leipzig, Herrmann estate)

They had flown poleward over the region between 11°30' W and 20°E, taken 11,600 photographs, and thus covered an area of 350,000 square kilometers and additionally inspected a further 250,000 square kilometers; therefore, a total area of approximately 600,000 square kilometers, named Neu-Schwabenland, had been explored.(Fig. 44)

In the end, a small boat excursion took place the next morning, during which 25 men were ferried to the pack ice. They got back on board with four shot seals, four more Adélie penguins, and several ice samples for the geophysicist.

On their way back home, regular oceanographic serial measurements of temperature and salt content, down to a depth of 6000 meters, commenced along the zero meridian.[90] At the same time, the biologist regularly lowered his net into the

---

[90] Paulsen, Karl-Heinz: Die ozeanographischen Arbeiten. in: Vorbericht über die Deutsche Antarktische Expedition 1938/39, in: Annalen der Hydrographie und Maritimen Meteorologie, VIII, Supplement 1939, pp. 28–32; Barkley, Erich: Die biologischen Arbeiten, in: Vorbericht über die Deutsche Antarktische Expedition 1938/39, Annalen der Hydrographie und Maritimen Meteorologie, VIII. Supplement, pp. 19–21; Ritscher: Wissenschaftliche und fliegerische Ergebnisse, pp. 99–105.

Fig. 42 The so-called Breischüsseln (now known as the Grautskåla Cirque) on the western border of the Humboldt Mountain Range [Alexander-von-Humboldt-Gebirge]. (Leibniz-Institut für Länderkunde, Leipzig, Herrmann estate)

Fig. 43 The Boreasic Lake District, subsequently renamed the Schirmacher Oasis. (Leibniz-Institut für Länderkunde, Leipzig, Ritscher estate)

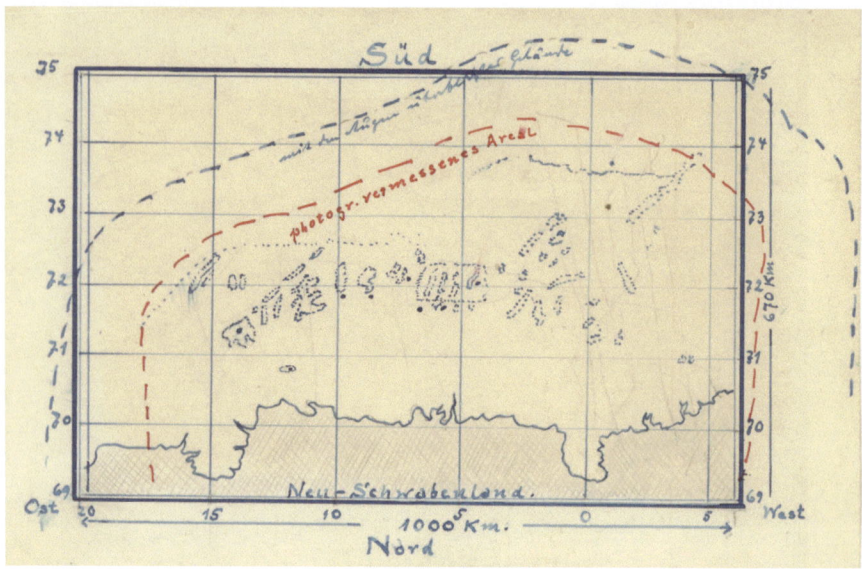

Fig. 44 Photogrammetrically scanned area with outlined locations of the discovered mountains and the inspected region of Neu-Schwabenland. (Leibniz-Institut für Länderkunde, Leipzig, Ritscher estate)

deep and the meteorologists followed the development of the weather in the high atmosphere with their radiosondes. West of Bouvet Island, Herrmann had the opportunity to survey the offshore bank they had discovered on their outward voyage in closer detail, by echo sounding.[91] Entertainment provided by the geophysicist Gburek dispelled the boredom of the hour-long waiting periods associated with the station measurements.[92]

On the occasion of 11,111 nautical miles traveled, an additional ration of alcohol was issued in the shape of a glass of schnapps or a bottle of beer.

After a sea voyage of nearly three months, the ship entered the port of Cape Town between March 6 and 7,[93] where Ritscher received a congratulatory telegram from Göring, praising the achievement as "worthy of Greater Germany's position in the world."[94] According to Wohlthat, the foundations were thus laid for acquisition of unowned land pursuant to international law, which meant expressing the will to occupy the land and permanently take over its administration and governance.[95]

When the *Schwabenland* departed from Cape Town, after bunkering drinking water and fresh food, Ritscher was eager to reach Recife by March 24 in order to

---

[91] Stocks, Theodor: Lotarbeiten der »Schwabenland« December 1938–April 1939, in: Vorbericht über die Deutsche Antarktische Expedition 1938/39, Annalen der Hydrographie und Maritimen Meteorologie, VIII. Supplement 1939, p. 37; Herrmann: Deutsche Forscher, pp. 111–113.
[92] Ibidem, pp. 120–122, 115.
[93] Ritscher: Wissenschaftliche und fliegerische Ergebnisse, pp. 107–114.
[94] Ibidem, p. 110.
[95] Wohlthat: Die Deutsche Antarktische Expedition 1938/39, pp. 616–617.

give the courier mail service of German Lufthansa a list of the names of the discovered mountain ranges and summits, among which were names such as Hermann-Göring-Land" (today: Neu-Schwabenland) or "Gralsburg" (today Trollslottet in Filchner Mountains).[96]

Then the secret exploration of the Brazilian Island of Trinidade, lying approximately 1200 kilometers west off the coast of Brazil, was put on the agenda.

Herrmann submitted a reported, in general terms, about having visited an island (meaning the island of Trinidade) that was so small "that it hardly had a name."[97] First, the high surf had to be overcome, during which the rowboat capsized, but everyone managed to reach the shore and, wet as they were, could explore the island. Ritscher, Lange, Herrmann, Barkley, and two seamen encountered deserted corrugated-iron huts, near which a small creek flowed. Then they climbed over steep, rough volcanic rocks in order to get to the next bay, where two rucksacks with food had been brought in a rubber dinghy. After they had spent the night under a bright, starry sky, they were picked up again. Ritscher mentioned in his travel report only that they should do some outboard work on the ship in the safety of the lonely and uninhabited island of Trinidade, which lay "in the middle of the ocean, alone in the world at 20°30'S, 29°00'W."[98]

The *Schwabenland* anchored in Cochoeiro Bay, which a Norwegian whale factory ship and several of her catcher boats entered shortly thereafter to take up fuel from the mothership in the wind shelter of the bay. On board the *Schwabenland*, pleasure was taken in shooting sharks and angling wonderfully multicolored fish, which, however, the biologist assessed as being poisonous, so they could not even be fed to the penguins. Ritscher reported to the high command of the German War Navy that a landing attempt on the southwest of the island on March 18 had not been possible, because of the steep cliffs.[99] Although Principe Bay had a 20- to 25-meter-wide fairway, its strong surf meant that it probably could not be navigated without difficulty, unless the sea was extremely quiet. In general, according to their experience, a potential landing on the northeastern side with dinghies, rubber boats, and experienced staff would invariably be risky. Of the decrepit huts, which had been built by Brazilian occupation troops during World War I, only four could still be used. However, the goats and pigs they had previously brought along had reproduced in the meantime. At night, giant turtles would come to the beach to lay eggs. A colonization of the island with perhaps 20 self-sustaining families would be quite imaginable, but the island would be frequented by whale factory ships more often than originally anticipated. The inspection of the northern part of the island would have to follow later, along with exploration of the neighboring island of Martin Vaz.

---

[96] Ritscher: Wissenschaftliche und fliegerische Ergebnisse, p. 111.
[97] Herrmann: Deutsche Forscher, pp. 135–141.
[98] Ritscher: Wissenschaftliche und fliegerische Ergebnisse, p. 112.
[99] Ritscher, Alfred: May 2, 1939, Brief an das Obekommando der Kriegsmarine, Leibniz-Institut für Länderkunde, Leipzig, Ritscher estate, File Bb1, Abt. Trinidad.

Fig. 45 The expedition members on the day of their homecoming, April 11, 1939. (Leibniz-Institut für Länderkunde, Leipzig, Ritscher estate)

To avoid paying harbor fees, the *Schwabenland* dropped anchor in the outer harbor of Recife for one night. Although abundant prophylactics against venereal diseases were handed out prior to each shore leave, one case of gonorrhea occurred after this short stay.[100] In addition, remedies had to be applied recurrently in cases of previous syphilis infections. After sufficient antiscurvy agents had been administered on the voyage, cases of vitamin deficiency no longer occurred during the expedition.

Upon leaving Recife, the ship took the fastest course homeward, via Fernando Noronha and Tenerife, and through the English Channel.[101] On April 11, 1939, the homecomers were given a ceremonious welcome by Helmuth Wohlthat and the President of the German Research Foundation, Rudolf Mentzel, in Cuxhaven (Fig. 45). They then continued on to Hamburg—with the engines running on full against the tidal outflow of the Elbe—where, at 7:00 p.m., a detachment of storm troopers [Sturmabteilung (SA)] was lined up in their honor. Mayor Carl Vincent Kroogmann welcomed the crew in the name of the "imperial governor" [Reichstatthalter] and gauleiter, Karl Kaufmann, in the great reception room at the City Hall. At the invitation of the Nazi Reich minister of science, education, and national culture [Reichsminister für Wissenschaft, Erziehung und Volksbildung], Bernhard Rust, the expedition was officially terminated at a banquet held at the Hotel Vier Jahreszeiten

---

[100] Bludau, Josef: (undated), Ärztlicher Bericht, Leibniz-Institut für Länderkunde, Leipzig, Ritscher estate, File Bh1, Abt. Medizin.

[101] Ritscher: Wissenschaftliche und fliegerische Ergebnisse, pp. 112–113. Herrmann: Deutsche Forscher, pp. 180–181.

Fig. 46 The medal of the third German Antarctic Expedition (1938–1939): rear side. (Hartmann, Gertraude (Private possession))

Fig. 47 The medal of the third German Antarctic Expedition (1938–1939): front side. (Hartmann, Gertraude (Private possession))

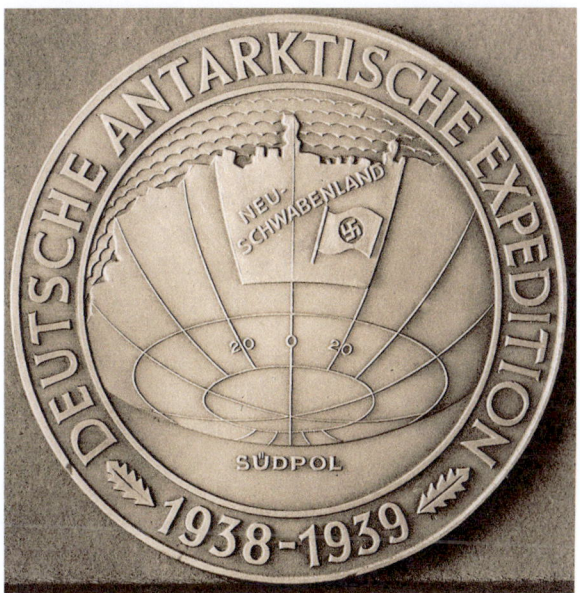

on the Binnenalster in Hamburg. For their achievements, the attendants were each awarded a medal (Figs. 46 and 47).

Ritscher, speaking in the name of the expedition, was "proud of having contributed to the fact that Germany has to be taken into account when it comes to solving the questions resulting from the work of our expedition in the field of international whaling in Antarctica, as regulated by the London Agreement, the regulation of the sovereign rights on this continent or the scientific research conducted there."[102] Like the previous German expeditions, these explorers had made an essential contribution to these ends.

The Evaluation, New Plans, and Untenable Myths

A preliminary report on the time course of the expedition and the preliminary results of the scientific work was released as early as 1939.[103] A map of Neu-Schwabenland (at a scale of 1:1,500,000) and two detailed maps of the western boundary of Alexander-von-Humboldt Mountain and the Wohlthat Massif (at a scale of 1:50,000) were of particular significance.[104]

The aerial photogrammetric images and the drawings made during the flights were the foundation for the construction of several ice coverage profiles in the study area of Neu-Schwabenland and enabled a large-surface analysis of one part of Antarctica's inland ice cap.[105] However, since no exact astronomical position findings of landmarks had been made, the coordinates of these sites on the map were fraught with great uncertainty and the map floated in midair, so to speak (Fig. 48).[106]

After the German ship *Meteor* and the British ship *Discovery II*, the *Schwabenland* was the third ship to survey the ocean floor by means of echo sounding. The *Schwabenland* recorded the first profile along a midoceanic ridge and displayed its rugged nature.[107]

On the basis of the high-resolution soundings made during the voyage of the *Schwabenland*, the geographer Herrmann was apparently the first person to interpret the suboceanic elevation referred to as the Atlantic Rise (now known as the Mid-Atlantic Ridge) in the South Atlantic as a volcanic phenomenon. It may

---

[102] Ritscher, Alfred: Neuland in der Antarktis, in: Münchner Neueste Nachrichten, issue of May 5, 1939.

[103] Vorbericht: Vorbericht über die Deutsche Antarktische Expedition 1938/39, in: Annalen der Hydrographie und Maritimen Meteorologie. VIII. Supplement 1939.

[104] Gruber, Otto von: Das Wohlthat-Massiv im Kartenbild, in: Ritscher: Wissenschaftliche und fliegerische Ergebnisse, pp. 157–230.

[105] Herrmann, Ernst: Die geographischen Arbeiten, in: Ritscher: Wissenschaftliche und fliegerische Ergebnisse, pp. 290–299; Klebelsberg, Raimund von: Formen und gletscherkundliche Auswertung der Lichtbildaufnahmen, in: Ritscher: Wissenschaftliche und fliegerische Ergebnisse, pp. 126–156.

[106] Kosack, Hans-Peter: Die Neubearbeitung der Übersichtskarte des Arbeitsgebietes der Expedition, in: Wissenschaftliche Ergebnisse Vol. 2, Issue 1, Hamburg 1954, pp. 1–15.

[107] Schumacher, Arnold: Die Lotungen der »Schwabenland«, in: Ritscher, Alfred (ed.): Deutsche Antarktische Expedition 1938/39, Wissenschaftliche und fliegerische Ergebnisse, Vol. 2, Hamburg 1958, pp. 41–62.

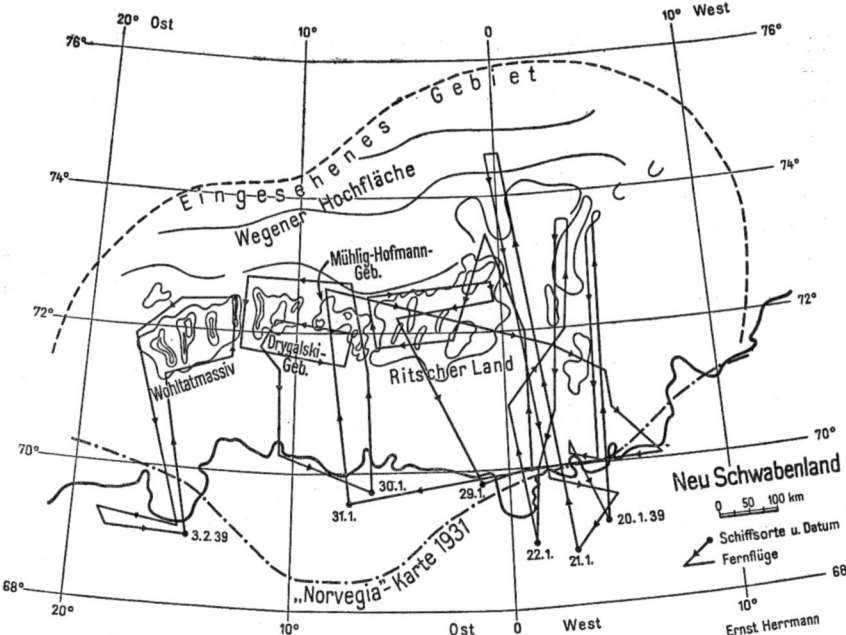

Fig. 48 Overview map of the ship-launched long-distance flights over Neu-Schwabenland. (Herrmann, Ernst: Deutsche Forscher im Südpolarmeer, Berlin 1941, p. 429)

"perhaps be nothing but a tremendous linear volcano, virtually a volcanic crest, only measuring a length that no one has ever observed before on our earth."[108] Subsequent studies revealed that the ocean floor was spread and pushed outward by the rising and cooling lava in the middle of the ridge, so the oldest material was to be found most far away. Although Herrmann was right about the volcanic origin of the Mid-Atlantic Ridge, his theory did not become known, as he initially published it in his travel report in 1941. His professional article appeared only in 1948 but was not taken notice of in English-speaking countries, and nor was the evaluation of the echo-sounding data, which was published as late as 1958.[109]

On April 12, 1939—hence, one day after the return of the expedition—Wohlthat gave the go-ahead for the execution of another expedition, which, in the next southern summer of 1939–1940, was supposed to lead to the Pacific sector between 80°W and 130°W, the territorial ownership of which was hitherto not in the least bit clear.[110]

---

[108] Herrmann: Deutsche Forscher, p. 165.

[109] Herrmann, Ernst: Tektonik und Vulkanismus in der Antarktis und den benachbarten Meersteilen, in: Petermanns Geographische Mitteilungen 1948, pp. 1–11; Schumacher, Die Lotungen.

[110] Wohlthat, Helmuth: June 6, 1939, W XVII/596, Leibniz-Institut für Länderkunde, Leipzig, Ritscher estate, File Bh1, Abt. VJP; see also Drygalski, Erich von: Entdeckungen und Ansprüche in der Antarktis, in. Geographische Zeitschrift 50 (1944) 1/2, p. 60.

Again, the primary objective was the drawing of a map. Although the Pacific sector had been declared a sanctuary at the London Whaling Conference in 1938, the occurrence of whales in the coastal waters was to be additionally examined.[111] This also included measurement of ocean currents and water depths.[112] A third objective concerned the ice shelf coast and its fauna. Of particular interest was determination of potential landing sites for later sledge expeditions. Exploration of the islands Trinidade and Martin Vaz was again planned to take place on the way back. The authority over the utilization of the current expedition results by the press was assigned to the Ministry of Propaganda, whereas the sovereign rights in Antarctica were to be addressed within the scope of political debates at a later event. However, further-reaching propagandist exploitation of the expedition did not take place.

Early in 1939, Ritscher submitted to Wohlthat a program draft for the following German Antarctic Expedition (DAE 39/40), including a cost estimation, and specified the expenses in a further letter.[113] Considering the daylight conditions, the outward voyage was to proceed, at the latest, on November 15, 1939, so that the expedition work could commence and go on around the clock by December 20. The overall duration of the voyage was planned to be five months, of which the stay in Antarctic waters was to cover six weeks. However, because of the outbreak of war on September 1, 1939, the execution of DAE 39/40 was cancelled at short notice, although "the *Schwabenland* in the harbor had already been put under steam,"[114] and the Office of the Four-Year Plan requested a final report on the previous activities.[115] Nevertheless, Wohlthat hoped for more profound consolidation of the commenced work and had Ritscher begin the new preparations in secrecy, apart from releasing the results of the previous expedition. Ritscher therefore drafted a second expedition program for DAE 40/41.[116] On the assumptions that the war would probably be over by the autumn of 1940 and that the time needed for making preparations would be extremely short, all previous German working areas were to be revisited in order to reassert territorial claims on Neu-Schwabenland and/or Kaiser

---

[111] Wohlthat, Helmuth: August 11, 1939, W XVII/742, Leibniz-Institut für Länderkunde, Leipzig, Ritscher estate, File Bh1, Abt. VJP.

[112] Wohlthat: June 6, 1939.

[113] Deutsche Antarktische Expedition: June 8, 1939, Entwurf eines Arbeitsprogramms der DAE 1939/40, Leibniz-Institut für Länderkunde, Leipzig, Ritscher estate, File Bh1, Abt. VJP; Deutsche Antarktische Expedition: August 24, 1939, Leibniz-Institut für Länderkunde, Leipzig, Ritscher estate, File Bh1, Abt. OKM.

[114] Conversation of Cornelia Lüdecke with Ilse Ritscher during the International Polar Conference in Gottingen on April 12, 1991.

[115] Wohlthat, Helmuth: September 5, 1939, Protokoll der Besprechung vom 5. September 1939, W XVII/797, Leibniz-Institut für Länderkunde, Leipzig, Ritscher estate, File Bh1, Abt. VJP.

[116] Ritscher, Alfred: August 19, 1940, Konzept, Leibniz-Institut für Länderkunde, Leipzig, Ritscher estate, File VAE1, Abt. no title; Ritscher, Alfred: August 26, 1940, Arbeitsprogramm für die DAE 1940–41, Leibniz-Institut für Länderkunde, Leipzig, Ritscher estate, File VAE1, Abt. no title, Ritscher, Alfred: September 14, 1940, Arbeitsprogramm für die DAE 1940–41, Leibniz-Institut für Länderkunde, Leipzig, Ritscher estate, File VAE1, Abt. no title.

Wilhelm II Land. Ritscher waited for the end of the war in vain.[117] In the meantime, the *Schwabenland* was chartered by the German War Navy.[118] In addition, in early January 1941, Ritscher himself had to report as a captain for war service in the English Channel.[119] At the end of October 1941, Ritscher put the matter to rest for good and closed the expedition office, until further notice, by November 1.[120] In January 1942, the stored polar expedition clothing was handed over to the German Army for use on the Eastern Front, subject to the condition that equivalent items of equipment would be manufactured as replacements for the expedition after the war, at the expense of the German Army. Finally, in August 1942, in the middle of World War II, the first part of the results of DAE 38/39 was published.[121] When the expedition ship *Schwabenland* was torpedoed and ran aground near the Norwegian town of Egersund in April 1944, all further expedition plans had become pointless.[122]

On July 10, 1945, two months after the end of World War II, two German submarines were sighted in the Argentinian naval port of Mar del Plata.[123] The Hungarian journalist Ladislas Szabó, who was living in exile in Argentina, reported this event under the title "Hitler in the Ice of Antarctica."[124] He thus founded a rapidly spreading myth, which has survived until today. In 1947, his sensational book *Je sais que Hitler est vivant* [*I Know that Hitler Is Alive*], including further elaborations to support his hypothesis, was published in Paris.[125] Allegedly, the *Schwabenland* expedition had installed a base for Hitler's retreat in Neu-Schwabenland (now known as Dronning Maud Land), consisting of gigantic caves in mountain rock, in which both airplanes and submarines could be housed and manned with a pertinent staff. Important key figures in the Third Reich had supposedly retreated there to prepare the return of the National Socialists. Even "flying saucers" were supposed to operate from this base.

---

[117] Ritscher, Alfred: September 26, 1940, Notiz, Leibniz-Institut für Länderkunde, Leipzig, Ritscher estate, File AK 2, Abt. G; Ritscher, Alfred: October 30, 1940, Note, Leibniz-Institut für Länderkunde, Leipzig, Ritscher estate, File AK 2, Abt. K.

[118] Deutsche Lufthansa: November 1. 1940, Deutsche Lufthansa, Leibniz-Institut für Länderkunde, Leipzig, Ritscher estate, File Bh2, Abt. VJP.

[119] Ritscher, Alfred: December 27, 1941, Notiz, Leibniz-Institut für Länderkunde, Leipzig, Ritscher estate, File AK 2, Abt. G.

[120] Ritscher, Alfred: October 2, 1941, Brief an Klebelsberg, Leibniz-Institut für Länderkunde, Leipzig, Ritscher estate, File AK 2, Abt. K.

[121] Ritscher: Wissenschaftliche und fliegerische Ergebnisse; see also the review in Drygalski, Erich von: Buchbesprechung von A. Ritscher (1942), in: Geographische Zeitschrift 49 (1943) 6, p. 284.

[122] Ritscher, Alfred: Vor 10 Jahren, in: Polarforschung 18 (1948) 1/2, p. 31, Lüdecke, Cornelia/ Colin Summerhayes: The Third Reich in Antarctica: The German Antarctic Expedition 1938–39, Eccles and Bluntisham, 2012, p. 92 and figure between 144 and 145.

[123] Summerhayes, Colin P./Peter Beeching: Hitler's Antarctic base: the myth and the reality, in: Polar Record 43 (2007) 224, p. 3.

[124] Szabo, Ladislav, Hitler esta vivo. El Tabano, Buenos Aires, 1947.

[125] Szabo, Ladislav, Je sais que Hitler est vivant. SFELT, Paris, 1947; Schön, Heinz: Mythos Neu-Schwabenland. Für Hitler am Südpol, Selent. 2004, pp. 145–155; Summerhayes/Beeching: Hitler's Antarctic base; Lüdecke/Summerhayes: The Third Reich in Antarctica, pp. 103–111.

How could such a myth be created? On the occasion of the largest American navy operation, Highjump, led by Admiral Richard Evelyn Byrd and comprising 13 navy ships and several aircraft, 4700 soldiers were to test their polar equipment in the southern summer of 1946–1947 and gain polar experience in Antarctica.[126] According to another interpretation, however, it served the purpose of seeking and destroying the German polar base. However, Byrd did not attach any importance to visiting Neu-Schwabenland whatsoever.[127]

It is a fact that in 1945, two submarines had actually fled to German-friendly Argentina to evade the need to surrender to the Americans. Their voyage to the south proceeded rather slowly, because fuel was lacking for this unforeseeably long trip and the submarines had to submerge in the daytime to escape discovery. After the capitulation, the time factor alone would not have allowed two submarines to make the long voyage to Argentina via Antarctica to build a Nazi base there. It was also overlooked that deepest winter prevailed in Antarctica during the northern summer of 1945, normally accompanied by a 1000-kilometer-wide and 1- to 2-meter-thick belt of sea ice surrounding the Antarctic continent completely. The ice fields piling up huge pack ice ridges could be conquered only with a lot of time and effort on the part of powerful icebreakers, and certainly not by submarines diving underneath all the way to the Antarctic coast without a reinforced hull or a sufficient external supply of air. Even the submarine used by Byrd's expedition, the *USS Sennet*, was damaged by the pack ice in the southern summer of 1946–1947 and had to be towed away.

Apart from that, the *Schwabenland*, which stayed at the Antarctic coast for only three weeks, had no loading capacity for additional caterpillar tractors, building material, food, or construction workers to build such a base in the mountains 200 kilometers away from the coast.[128] Besides, building a station in this region would have been a huge logistical task. First, the building material would have had to be moved to the inland mountains, then tunnels would have had to be excavated in the rock with dynamite, with the boulders transported off in trucks on rails, as occurs in ordinary mining operations. Whether the base would have been a habitable place for a few people at the end of the season basically remained an open question. And why would all of this have happened with so much foresight early in 1939 in order to enable an escape of Nazi leaders six years later?

Another myth holds that a nuclear bomb was even detonated in 1958 in order to destroy the German base.[129] However, by this time, all of the people had already left

---

[126] Headland: Chronology, pp. 313–314.

[127] Rose, Lisle A.: Assault on Eternity. Richard E. Byrd and the Exploration of Antarctica 1946–47, Annapolis, 1980, p. 225; Godwin, Joscely: Arktos. The Polar Myth in Science, Symbolism, and Nazi Survival, London, 1996, pp. 125–129; Lüdecke/Summerhayes: The Third Reich in Antarctica, pp. 104–109.

[128] Interview with Siegfried Sauter in Schön: Mythos Neu-Schwabenland, pp. 160–168; Lüdecke/Summerhayes: The Third Reich in Antarctica, pp. 105–106.

[129] Summerhayes/Beeching: Hitler's Antarctic base; Lüdecke/Summerhayes: The Third Reich in Antarctica, pp. 109–111.

for South America. In the scope of an experiment during the International Geophysical Year (1957–1958), three nuclear bombs had, in fact, been detonated over the South Atlantic at an altitude of between 160 and 750 kilometers in order to find out whether the loaded particles and radioactive isotopes released during the detonation interacted with the earth's magnetic field. However, this had happened about 2300 kilometers off the coast of Neu-Schwabenland. Besides, a search for fallout from all nuclear bombs proved that there had never been a nuclear attack on Antarctica.

The aforementioned myth about unidentified flying objects (UFOs) or flying saucers or German secret weapons in connection with a German retreat base in Antarctica is without any foundation and resulted from a translation error made in a Spanish-language article published in the Chilean *El Mercurio* on March 5, 1947, which, when translated correctly, stated that Byrd had warned the USA against an "invasion of the country by hostile aircraft which would be coming from the polar region."[130] However, in an embellished English translation, it said that "in case of another war the continental United States would be attacked by flying objects which could fly unbelievably fast from pole to pole."[131]

Despite the fact that the aforementioned myths lacked any rational foundation whatsoever, they survive in the minds of their supporters until today.

Even during student parliament elections for the legislative period of 2015–2016, the Alternative Linke [Alternative Left]—which advocates nonpartisan, grassroots democracy and an autonomous university—displayed posters on the campus of the University of Hamburg, drawing attention to a map of Antarctica. A line separated the upper part of Antarctica from the rest of the continent, and an arrow pointing to it said "Nazis to Neuschwabenland" (Fig. 49).

Excursion: Elements of German Antarctic Expeditions Before World War II

All elements characterizing today's polar research were developed in the course of the first three German Antarctic expeditions: oceanographic studies on the outward and return voyages, establishment of an overwintering station in Antarctica, and short summer campaigns with aircraft. However, the organizational form of the individual expeditions differed considerably.

In about 1900, the first German South Polar Expedition was financed by the state because of its political significance in a time when the world was undergoing divisions. Erich von Drygalski was appointed as the leader of the expedition at an early stage and made preparations for it with the aid of advisory commissions and self-chosen expedition members, for whom even the first German polar expedition ship (the *Gauss*) was built. He was also the empire's representative on board and legally

---

[130] Lüdecke/Summerhayes: The Third Reich in Antarctica, p. 109.
[131] Ibidem.

Fig. 49 Canvassing poster for the Alternative Linke during student parliament elections, as seen on the campus of the University of Hamburg on January 16, 2015. (Lüdecke, Cornelia (Private possession))

outranked even the captain. However, after the expedition returned without politically exploitable results, the *Gauss* was sold to a foreign country, thus preventing the continuation of German polar research.

Ten years later, Wilhelm Filchner could no longer hope for state financing. For that reason, he collected the money needed for his privately organized expedition and the ship *Deutschland* through a lottery (Fig. 50). The funds raised were administered by an association. In addition, this association, which employed all expedition members (including Filchner himself), took over the further preparations. The captain represented the law on board and thus outranked Filchner. As is still customary on expeditions today, in Filchner's time there was a so-called cruise leg on the outward journey from Hamburg to Pernambuco, for which a "cruise leader" (Filchner's deputy) had responsibility. Huge interpersonal problems arose on the actual voyage to Antarctica, as Filchner himself had not joined the expedition crew until they reached South America, after separate factions had already formed on the *Deutschland*.

The third Antarctic expedition was planned only as a short summer campaign and was organized by ministries in such a manner that the desired results would be attained by efficient application of all available means. All threads led back to Helmuth Wohlthat in the Office of the Four-Year Plan, whereas the technical

Excursion: Elements of German Antarctic Expeditions Before World War II 191

Fig. 50 The *Deutschland*, frozen in ice, during the second German Antarctic Expedition in 1912. (Joester, Erich (Private possession))

support came from other ministries and state boards, which also selected the participants and provided the equipment (the ship and aircraft). As an expedition leader, Alfred Ritscher did not have much freedom, for he had to fulfill the defined expedition plan within the given time frame. While the expedition was in progress, the participants developed a research flight routine that is still generally applied today[132]:

1. Discussion in the evening:
    - Weather forecast for the next day (Fig. 51)
    - Planning of the flight for the next day

2. Discussion and decision on the flight route:
    - Determination of the takeoff time

3. Immediately before the research flight:
    - Short-term weather forecast ("nowcast")
    - Pilot's briefing

4. Research flight

5. After the research flight:

---

[132] Lüdecke, Cornelia: Investigation of the unknown: The flight programme of the German "Schwabenland" expedition 1938/39, in: The Polar Journal 2 (2012) 2, pp. 312–333.

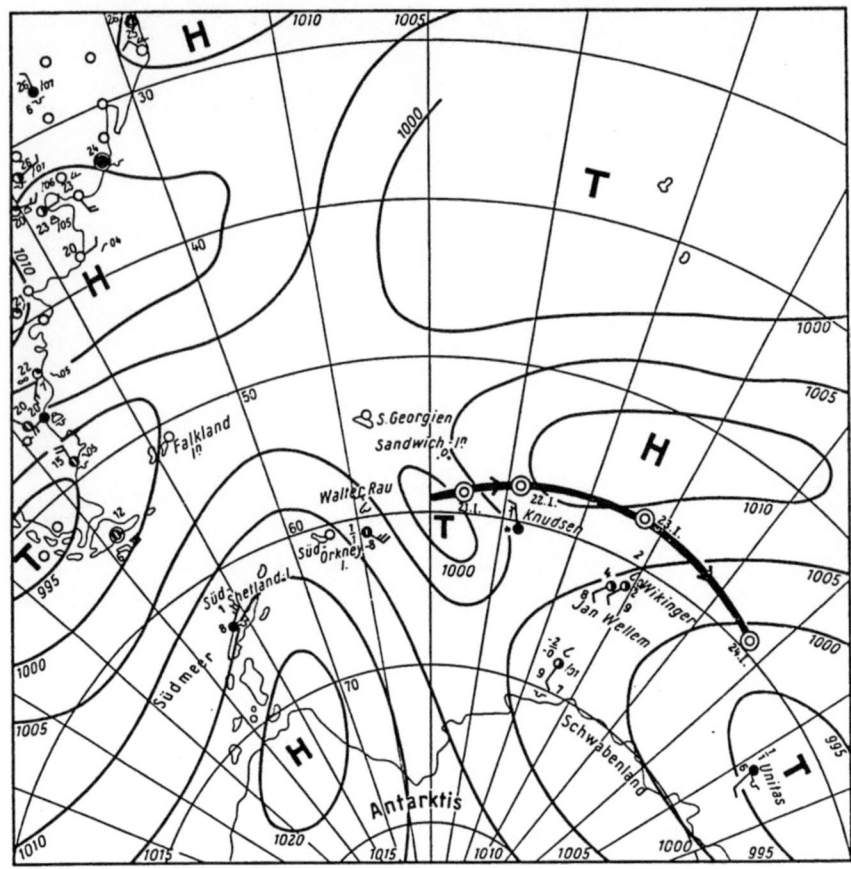

Fig. 51 Surface weather map from January 20, 1939, and the path of a depression from January 20 to 24, 1939, providing the basis for a weather forecast on board. (Regula, Herbert: Die Wetterverhältnisse während der Expedition und die Ergebnisse der meteorologischen Messungen, in: Ritscher, Alfred (Hg.): Deutsche Antarktische Expedition 1938/39. Wissenschaftliche Ergebnisse Vol. 2 Issue 1, Hamburg 1958, p. 2)

- Maintenance of the aircraft
- Repair of equipment

6. Discussion in the evening:

- Oral flight report
- Oral technical report
- Submission of notes and drawings made on the flight
- Submission of a written flight report (at the earliest, one day afterward)
- Planning of the next flight

Separate and United Paths: German Antarctic Research from the End of World War II Until Today

A Private Initiative to Resume Antarctic Research in the Federal Republic of Germany

A German territorial claim on Neu-Schwabenland has never been registered. The beginning of World War II interrupted the history of Germans in Antarctica. Exploration of the South Polar region was of no avail in military confrontations. In 1952, the state secretary in the Foreign Office of the Federal Republic of Germany (FRG; West Germany), Walter Peter Hallstein, officially published the names of the discoveries made during the Antarctic expedition of 1938–1939 in the *Federal Gazette* and announced that coordinated meteorological, magnetic, and further geophysical measurements would be in a state of planning at over 30 overwintering stations.[1] This happened in the run-up to the International Geophysical Year, which lasted from July 1, 1957, to December 31, 1958 (and is now referred to as the Third International Polar Year). However, the FRG did not participate with an expedition. Instead, there was a private initiative by Karl Maria Herrligkoffer, a physician and leader of two Himalayan expeditions, to establish a station close to Byrd Bay in Neu-Schwabenland during the International Geophysical Year in order to continue the studies that Ritscher's expedition had begun (Fig. 1).[2]

In a brochure in which he advertised his Antarctic expedition, Herrligkoffer underscored the strategic and economic significance of Antarctica with regard to the mineral resources presumed to be there—for example, uranium. On the political level, he would be able to build the foundation of a permanently manned scientific

---
[1] Hallstein, Walter Peter: Bekanntmachung über die Bestätigung der bei der Entdeckung von »Neu-Schwabenland« im Atlantischen Sektor der Antarktis durch die Deutsche Antarktische Expedition 1938/39 erfolgten Benennungen geographischer Begriffe, in: Bundesanzeiger 4 (1952) 149, issue of August 5, 1952, pp. 1–2.
[2] Herrligkoffer, Karl Maria: Deutsche Südpol-Expedition, Munich 1956; Bruns, Herbert: Deutsche Südpol-Expedition 1956/58. Ein akutes Problem der Gegenwart, Munich 1956.

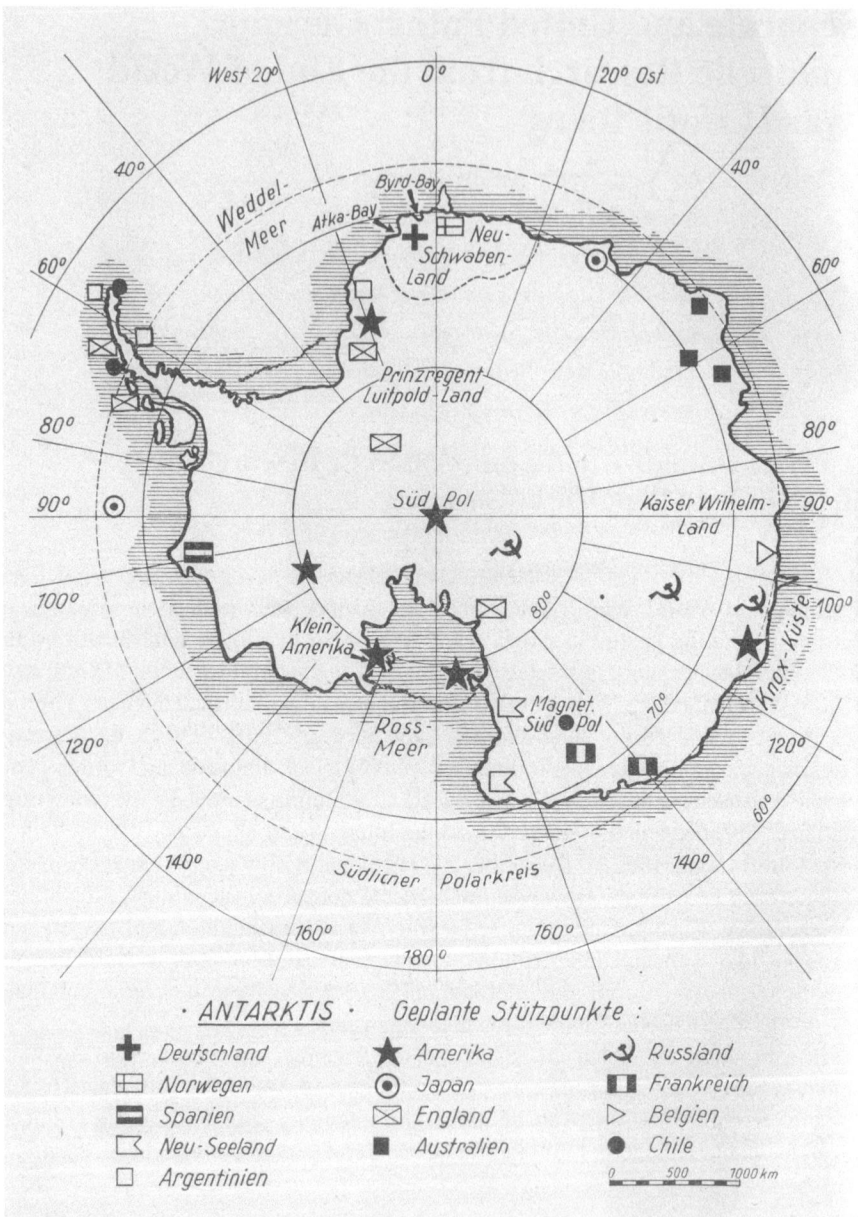

Fig. 1 A map of the Antarctic stations planned in the course of the International Geophysical Year (1957–1958). (Herrligkoffer, Karl Maria: Deutsche Südpol-Expedition, München 1956, p. 4)

station with his expedition and thus give Germany the opportunity to successfully assert territorial claims on Neu-Schwabenland before the Antarctic authorities during their planning when the International Geophysical Year was over. Herrligkoffer was alluding to the Scientific Committee on Antarctic Research (SCAR), which was founded in 1958.[3] But the geographers present at the 30th Conference of German Geographers in Hamburg vehemently spoke out against Herrligkoffer's proposal, as his overscaled expedition plan, foreseeing the participation of 30 scientists, appeared to be absolutely unrealistic.[4] In the end, the realization of Herrligkoffer's expedition failed in 1957 because of the question of money.

The International Geophysical Year (1957–1958) Without German Participation

During the International Geophysical Year (1957–1958), 12 nations established a total of 54 scientific stations: 15 stations on subantarctic islands and a total of 39 on the Antarctic coast, on the Antarctic ice cap, and at the South Pole itself.[5] In addition, the Commonwealth Trans-Antarctic Expedition, led by Vivian Fuchs, traversed Antarctica for the first time in the southern summer of 1957–1958 from the Weddell Sea to the Ross Sea, on a route similar to the one originally planned by Wilhelm Filchner. By applying the latest methods of engineering and measurement, among which were the first satellites, Antarctica was explored comprehensively from all sides.

After the International Geophysical Year, the 12 countries active in Antarctica adopted the so-called Antarctic Treaty on December 1, 1959, which secured peaceful international research and deferred all territorial claims on Antarctica asserted in the past.[6] On June 23, 1961, the signatory states (Argentina, Australia, Belgium, Chile, France, Great Britain, Japan, New Zealand, Norway, South Africa, the Soviet Union, and the USA) ratified the Antarctic Treaty and agreed on nonmilitary exploitation of Antarctica for scientific purposes.[7] The topics could be negotiated again only in 1991, i.e., 30 years after the treaty had come into force. The success of the Antarctic

---

[3] Headland: Chronology, pp. 29, 362; Walton, David W. H./Clarkson, Peter D.: Science in the Snow. Fifty years of international collaboration through the Scientific Committee on Antarctic Research, Cambridge 2011.

[4] Lüdecke, Cornelia: Karl Maria Herrligkoffer's private "German South Pole Expedition" 1957/58: A failed initiative, in: Lüdecke, Cornelia (ed.): Steps of Foundation of Institutionalized Antarctic Research. Proceedings of the 1st SCAR Workshop on the History of Antarctic Research, Munich 2–3 June 2005, in: Reports on Polar and Marine Research 560 (2007), p. 199.

[5] Headland: Chronology, pp. 349–350, 355–356.

[6] Ibidem, pp. 29–32.

[7] Kohnen, Heinz: Antarktis Expedition. Deutschlands neuer Vorstoß ins ewige Eis, Bergisch Gladbach 1981, pp. 202–206; the contract can be downloaded at: https://www.ats.aq/e/antarctic-treaty.html

Fig. 2 A group photo at the 2009 Antarctic Treaty Summit at the National Museum of Natural History in Washington, DC (USA). *From left to right*: Jim Barnes, Guillan Triggs, Scott Miller, Cornelia Lüdecke, Michael A. Lang, Paul Arthur Bergman, Olav Orheim, Nina Federoff, Prince Albert II of Monaco, G. Wayne Clough, Aant Elzinga, Marie Jacobsen, Michael Richardson, R. Tucker Sally, David Walton, Stephen Rintoul, and Jorge Berguño. (Bergman, Paul (Private possession))

Treaty and its survival were acknowledged on the occasion of its 50th anniversary at the Antarctic Treaty Summit in Washington, DC (USA), in 2009 (Fig. 2).[8]

Antarctic Research in the German Democratic Republic Since 1960

In 1956, the Council of Ministers [Ministerrat] of the German Democratic Republic (GDR) passed a resolution to participate in the International Geophysical Year, including research activities outside the country's own frontiers, but this commitment did not extend to Antarctica.[9] In 1958, the meteorologist Günter Skeib received an opportunity to lead a meteorological working group from the glaciological expedition of the Kazakh Academy of Sciences into the Tian Shan Mountains in Central

---

[8] Berkman, Paul Arthur/Lang, Michael A./Walton, David W. H./Young, Oran R. (ed.): Science Diplomacy: Antarctica, Science, and the Governance of International Spaces, Washington, DC, 2011.

[9] Paech: Die DDR-Antarktisforschung, pp. 198–190, 201.

Asia.[10] Skeib's contacts with Soviet glaciologists later resulted in an invitation for him to lead a group of three GDR meteorologists taking part in the fifth Soviet Antarctic Expedition (SAE 1959–1961) to Mirny Station.[11] After the Council of Ministers passed their resolution, the GDR's Antarctic research was initially organized by the National Committee for Geodesy and Geophysics [Nationalkomitee für Geodäsie und Geophysik].[12] Participation in Antarctic research was considered a national effort and served the purpose of national recognition. In particular, it was supposed to strengthen the influence of the socialist states in Antarctica.[13]

In 1960, Skeib overwintered together with two Russians in a small semispherical tent on the approximately 21-kilometer-long and 14-kilometer-wide Drygalski Island, where they carried out meteorological observations and radiation measurements.[14]

Three days before Skeib and his two companions returned to Mirny, a fire had broken out in the meteorological station during a terrible gale-force wind on the night of August 2–3, 1960, during which Christian Popp and seven Soviet scientists lost their lives.[15] Fire is one of the greatest dangers in polar regions, for snow is hardly of any use to extinguish it. Despite this tragic event, the overwintering episode was the beginning of a continual involvement of GDR scientists in Soviet expeditions. With the aid of scientific counseling and logistical support in the scope of Soviet Antarctic expeditions, the groups from the GDR fulfilled "the scientific and societal assignment conferred upon them with great commitment," according to an official representation.[16] In the southern winter of 1969, there was another fatality when a German scientist, Klaus Diederich, fell from an ice barrier on July 17 and died from internal injuries.[17]

In 1969, the organization of the GDR's Antarctic research was transferred from the National Committee for Geodesy and Geophysics to the Central Institute of Physics of the Earth [Zentralinstitut für Physik der Erde (ZIPE)] in Potsdam.[18] Year after year, GDR scientists conducted various studies in Antarctica. For example, on the 17th Soviet Antarctic Expedition (SAE 1971–1973), the geodesist Siegfried

---

[10] Fritzsche, Dietrich/Gernandt, Hartwig/Foken, Thomas: In Memoriam Günter Skeib, in: Polarforschung 81 (2011) 2, pp. 127–128; Meier, Siegfried: 450 Tage in Antarktika, Leipzig 1975, p. 236; Tripphahn, Bodo: Aus der Werkstatt der Expeditionen, in: Lange, Gerd: Sonne, Sturm und weiße Finsternis. Eine Chronik der ostdeutschen Antarktisforschung, Hamburg 1996, p. 277.

[11] Fritzsche/Gernandt/Foken: In Memoriam, p. 127; Hempel, Gotthilf: Forscher, die in die Kälte gehen, in: Die Zeit, issue of November 3, 1989.

[12] Paech: Die DDR-Antarktisforschung, p. 199; Tripphahn: Aus der Werkstatt, pp. 277–283.

[13] Fritzsche, Diedrich: Geowissenschaftliche Forschung der DDR in der Antarktis, in: Schriftenreihe für Geowissenschaften 18, Zur Geschichte der Geowissenschaften in der DDR—Part II, Ostklüne 2011, p. 303.

[14] Skeib, Günter: Orkane über Antarktika. Forscherarbeit in Schnee und Eis, Leipzig 1963, pp. 139–159.

[15] Skeib: Orkane über Antarktika, pp. 160–161.

[16] Lange, Gerd: Bewährung in Antarktika. Antarktisforschung der DDR, Leipzig 1982, pp. 214–215.

[17] 7th GDR Expedition Group—14th SAE: January 1, 2000/PolarNEWS, http://antarktis.ch/2000/01/01/7ddrexpeditionsgruppe14sae/

[18] Paech: Die DDR-Antarktisforschung, p. 199, Skeib: Orkane, p. 160.

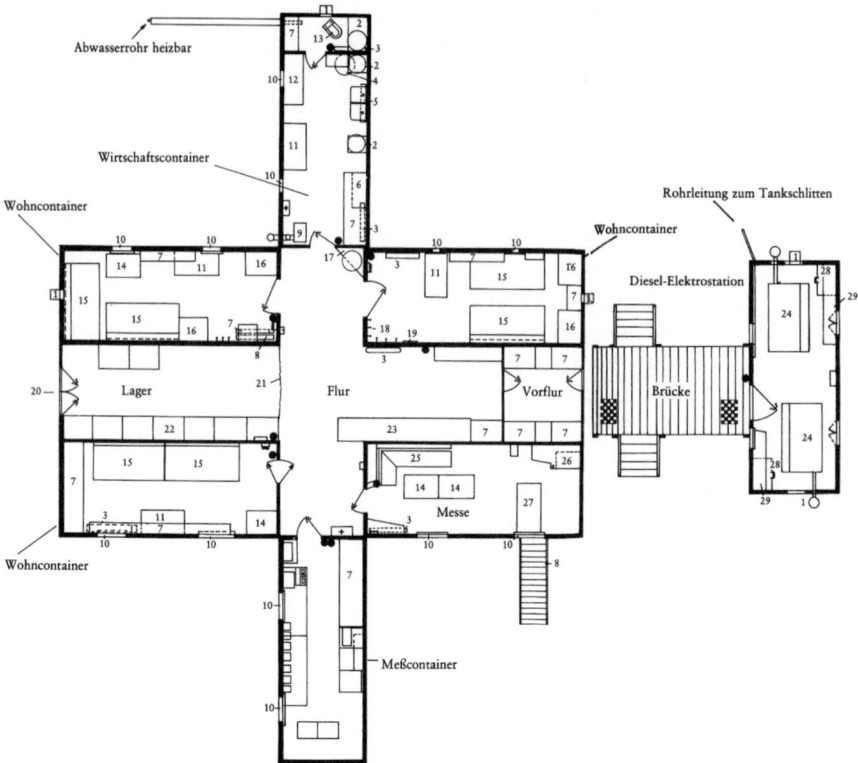

Fig. 3 A 1976 floor plan of the German Democratic Republic container station (later called Georg Forster Station) at the Schirmacher Oasis: *1* ventilator, *2* shelves for water containers, *3* radiators, *4* zinc barrel for snow melting, *5* washbasin, *6* photography laboratory, *7* shelves, *8* staircase to the roof, *9* oil furnace, *10* window, *11* table, *12* work table, *13* toilet, *14* camping table, *15* bed, *16* cupboard, *17* cupboard for food, *18* wardrobe, *19* mirror, *20* emergency exit, *21* curtain, *22* shelf rack, *23* table for instruments, *24* diesel generator, *25* bench, *26* satellite receiving system, *27* desk, *28* additional fuel tank (100 liters), *29* workbench, * fire extinguishers. (Gernandt, Hartwig; Erlebnis Antarktis, Berlin 1984, pp. 258–259)

Meier took part in a 120-kilometer-long geodetic–glaciological traverse from Molodezhnaya Station to the nunataks located in the southeast.[19]

Since no further financial demands were involved, the GDR joined the Antarctic Treaty in 1974.[20] In 1976, during the 21st Soviet Antarctic Expedition, Hartwig Gernandt led the largest GDR overwintering crew of six men, who built an independent base out of containers, providing 120 square meters of usable space at the Schirmacher Oasis, only 1.5 kilometers away from the Soviet Novolazarevskaya Station (Figs. 3 and 4). This was the foundation of the later Georg Foster Station.[21]

---

[19] Meier: 450 Tage in Antarktika, pp. 93–121, 126.

[20] Paech: Die DDR-Antarktisforschung, p. 198, 202. There were no financial means to establish a research station of one's own, so use of the advantages of a "lively German–Soviet friendship" was greatly appreciated; see also Lange: Bewährung in Antarktika, p. 215.

[21] Gernandt, Hartwig; Erlebnis Antarktis, Berlin 1984, pp. 232–270.

Fig. 4 The German Democratic Republic's George Forster Station, near the Soviet Novolazarevskaya Station at the Schirmacher Oasis, photographed in 1988. (Strecke, Volker (Private possession))

Originally, the main task was a comprehensive study of the ionosphere (at an altitude between approximately 70 kilometers and 400 kilometers), in which the polar lights appear.

On October 9, 1979, the Council of Ministers decided that the GDR should achieve consultative status under the Antarctic Treaty in order to participate in the conferences of the consultative countries (i.e., the original signatory countries of the Antarctic Treaty and further countries engaged in research in Antarctica), which, among other activities, were held to elaborate recommendations for research and logistics, and to review the efficacy of the treaty 30 years after adoption of the Antarctic Treaty.[22] However, the GDR could not realize the establishment and maintenance of a permanently manned Antarctic station, which was the basic requirement for such an application, soon enough.[23] Yet the membership application submitted by the GDR Academy of Sciences was officially approved in 1981 after the foundation of a National Committee for Antarctic Research, and the GDR was accepted as a member of SCAR.[24] Meanwhile, GDR scientists continued exploring the environment of the Schirmacher Oasis. During the 28th Soviet Antarctic Expedition, GDR scientists followed up on Alfred Ritscher's research when they

---

[22] Lange: Bewährung in Antarktika, p. 213.

[23] Paech: Die DDR-Antarktisforschung, pp. 198–203.

[24] Lange: Bewährung in Antarktika, p. 214.

investigated the frozen Lake Untersee in the Wohlthat Massif for the first time, early in December 1983.[25]

From 1985 to 1992, the distribution of ozone was regularly measured all the way up into the stratosphere with ozonesondes from the GDR base at the Schirmacher Oasis. These data later became particularly valuable when the so-called ozone hole was discovered by satellites, as only they could give an insight into changes in the ozone distribution at various altitudes over a period of seven years. Further main research topics were isotope measurements of snow and ice samples to get information about snow accumulation rates and temperature profiles during previous times.

The GDR policy was increasingly put under pressure by the Soviet Union, which, for many years, urged the GDR to establish an Antarctic station of its own.[26] Moreover, polar researchers from other nations did not understand why the GDR still kept running its base as an annex to the Soviet station and did not "operate its own station also in the sense of international law."[27] Hence, the research base at the Schirmacher Oasis became Georg Forster Station on October 25, 1987, and an overwintering crew was sent for the first time to an Antarctic station of their own, meaning that the GDR's first Antarctic expedition became a reality. In this context, it proved extremely valuable that logistical efforts were still managed by the Soviet Union.

Official bodies did not promote contacts with West German polar researchers in general or with the FRG's Georg von Neumayer Station in particular; however, an Antarctic exchange did exist in connection with weather messages or amateur radio communications (Fig. 5).[28]

When the crew members of the GDR's third Antarctic expedition set out to the south in the autumn of 1989, little did they know to what kind of a politically changed Germany they would return.[29]

---

[25] Simonov, I. M./Stackebrandt, W./Haendl, D./Kaup, E./Kämpf, H./Loopmann, A.: Report on Scientific Investigations at the Untersee and Obersee Lake, Central Dronning Maud Land (East Antarctica), in: Geodätische und geophysikalische Veröffentlichungen Reihe I (1985) Issue 12, p. 9; Fleischmann, Klaus: Zu den Kältepolen der Erde. 50 Jahre deutsche Polarforschung, Bielefeld 2005, pp. 131–134.

[26] Paech: Die DDR-Antarktisforschung, pp. 202–203, Fleischmann: Zu den Kältepolen, pp. 213–215.

[27] Lange: Sonne, Sturm und weiße Finsternis, pp. 233–237.

[28] Personal communication from Volker Strecke, Munich, January 28, 2015.

[29] Sobiesiak, Monika/Susanne Korhammer (ed.): Neun Forscherinnen im ewigen Eis. Die erste Antarktisüberwinterung eines Frauenteams, Basel, Boston, Berlin 1994, p. 117; Fleischmann: Zu den Kältepolen, p. 284.

Fig. 5 The radio communications container at Georg Forster Station, *left*: The official radion station used for land and sea communications (Y3ZA). *Right*: The amateur radio station Y88POL, photographed in 1991. (Strecke, Volker (Private possession))

Antarctic Research in the Federal Republic of Germany Since 1975

In the mid-1970s, the oil crisis of 1973 had just been overcome when the FRG turned again to Antarctica in search of raw materials and thematically resumed the German polar research that had stopped before World War II. First, commercial interest in fishery resources, especially in krill (as the largest protein reserve in the oceans), was of primary importance.[30] Instead of whale research, which had still been the central topic in 1938–1939, the first two West German Antarctic expeditions (in 1975–1976 and 1977–1978) focused on studying krill as the starting point of the food chain. On January 18, 1978, the federal government decided to accede to the Antarctic Treaty. It was decided that the FRG should become an "Antarctic power" as soon as possible and—equipped just like the Americans, Russians, and Japanese—should conduct Antarctic research to secure the provision of raw materials.[31] Apart from ore and valuable minerals, oil reserves were mentioned, which

---

[30] Kohnen: Antarktis Expedition, pp. 10–16, Mandelsloh, Klaus von/Freyenhagen, Jörn: Antarktis. Entdeckungsfahrten in die Zukunft, München 1984, pp. 63–79; Fleischmann: Zu den Kältepolen, pp. 184–190.

[31] Bergdoll, Udo: Noch ist alles eingefroren. Warum Minister Matthöfer darauf drängt, daß Bonn möglichst bald dem »antarktischen Club« beitritt, in: Süddeutsche Zeitung, issue of February 13, 1978.

were supposed to be greater than those in Alaska. By studying krill, Germany had "already put one foot in the Antarctic door."[32] Research Minister Hans Hermann Matthöfer proposed to "enter the race for Antarctic raw materials."[33] An initial investment of 90 million deutsche marks (DM 90 million) would be needed to establish a research station and build an ice-reinforced research vessel, a prospect that Matthöfer offered in his designated position as the finance minister (1978–1982). Consequently, in May 1978, the first step was taken by the members of the Antarctic Treaty and the German Research Foundation was accepted as a member of SCAR.

In the season of 1979–1980, a location for the West German Antarctic station on the Filchner–Ronne Ice Shelf was to be found, using the chartered Norwegian icebreaker *Polarsirkel*.[34] Two possible locations were determined in the course of the expedition: one on the southern border of the Weddell Sea near Wilhelm Filchner's research area before World War I, and another on the Ekström Ice Shelf near Atka Bay, close to the station site that had already been proposed by Herrligkoffer in the 1950s.[35] The initial planning for the polar research vessel *Polarstern* also began at this time.

The Federal Institute for Geosciences and Natural Resources [Bundesanstalt für Geowissenschaften und Rohstoffe (BGR)] also began to take an increasing interest in Antarctica and, in the season of 1979–1980, dispatched the first German Antarctic North Victoria Land Expedition (GANOVEX I) to the Ross Sea, on board the chartered icebreaking supply vessel *Schepelsturm*, to search for exploitable natural resources in North Victoria Land. One year earlier, nothing had been found in the Weddell Sea area.[36] In the background stood basic research for testing the Gondwana theory, which explains how the primordial great continent of Gondwana broke apart into multiple separate continents. On January 14, 1980, the Lillie Marleen Hut was erected near the Lillie Glacier in North Victoria Land as a land base for geological research programs. In 2005, it became the 79th entry in the list of Historic Sites and Monuments in Antarctica and was the first German facility to be listed (Fig. 6).[37]

Finally, on July 15, 1980, the Alfred Wegener Institute for Polar Research (AWI) was founded for coordination and logistical support for polar expeditions, maintenance of and communication with polar stations, and evaluation of comprehensive

---

[32] Bergdoll: Noch ist alles eingefroren.

[33] Ibidem.

[34] Kohnen: Antarktis Expedition, pp. 15–32, 45–47; Mandelsloh/Freyenhagen: Antarktis. pp. 70–79; Enß, Dietrich: Die deutsche Antarktis-Forschungsstation, in: Hansa 118 (1981) 13, p. 963.

[35] Anonymous: Antarktisforscher brachen Rekord, in: Hamburger Abendblatt, issue of March 10, 1980.

[36] Mandelsloh/Freyenhagen: Antarktis, pp. 91–115; Fleischmann: Zu den Kältepolen, pp. 172–180.

[37] www.bgr.bund.de/DE/Themen/Polarforschung/Antarktis/Logistik/logistik_node.html; www.ats.aq/documents/ATCM36/WW/atcm36_w w004_e.pdf

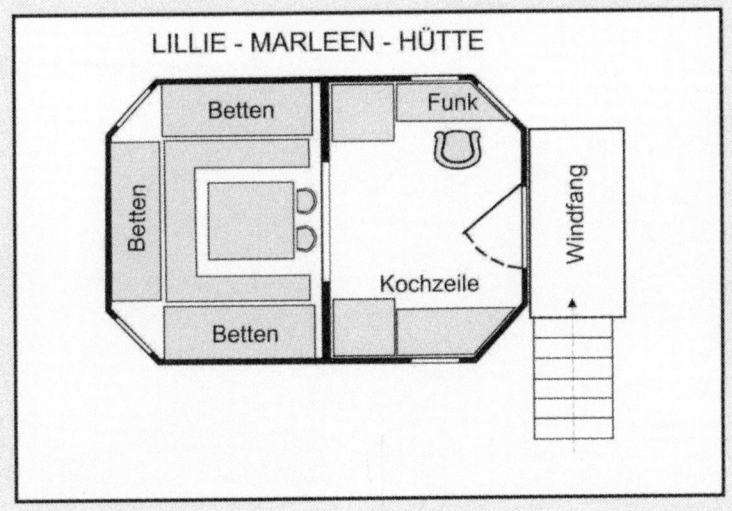

Fig. 6 The floor plan of the Lillie Marleen Hut, which serves as a base for geological field research in North Victoria Land. (Bundesanstalt für Geowissenschaften und Rohstoffe, Hannover)

measurements.[38] In 1980–1981, three ships were headed for the Weddell Sea at the same time to build Georg von Neumayer Station. The *Gotland II* carried all of the building materials. To everyone's surprise, however, dense pack ice blocked their access to the coast, and so they had to turn around, just 170 kilometers from their original destination, and approach the alternative location in Atka Bay (Figs. 7 and 8).

The time pressure was immense, for the West German government was waiting for a "success message that the station had been built"—a prerequisite for the impending consultative status requiring an Antarctic station to be in operation all year round.[39]

On January 23, 1981, unloading of the *Gotland II* began and the construction camp was set up (Fig. 9).[40] Floodlights to illuminate the construction site started to be installed early in February. In the first 20 days, work proceed in two shifts or sometimes even three shifts around the clock, in order to use the construction tools without stopping. Later, there were prolonged shifts of 10–12 hours. The station was built within a period of 33 days and consisted of tubes arranged in the shape of an *H*, housing 24 heated living rooms, work space, and supply containers (Figs. 10 and 11).

---

[38] Hempel, Gotthilf: Zum Aufbau des Alfred-Wegener-Instituts für Polarforschung, in: Polarforschung 51 (1981) 2, pp. 241–246; Fleischmann: Zu den Kältepolen, p. 218.

[39] Mandelsloh/Freyenhagen: Antarktis. p. 71.

[40] Personal communication from Friedrich Obleitner on February 23, 2015; personal communication from Dietrich Enss on August 28, 2015.

Fig. 7 The *Gotland II* cruising in Atka Bay after transporting material to Georg von Neumayer Station, photographed in 1981. (Albrecht, Klaus-Peter (Private possession))

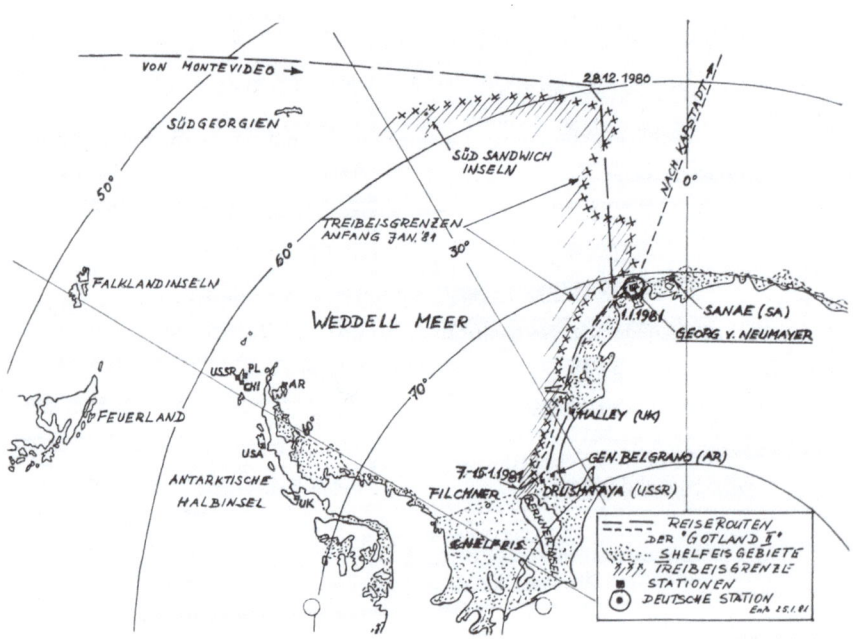

Fig. 8 The travel route of the *Gotland II* in the Weddell Sea and ice conditions in early January 1981. (Enß, Dietrich: Die deutsche Antarktis-Forschungsstation, in: Hansa 118 (1981) 13, p. 965)

Fig. 9 Diesel needed to run the generators at Georg von Neumayer Station is transported by helicopter for over seven kilometers onto the ice shelf in January 1981. (Albrecht, Klaus-Peter (Private possession))

Fig. 10 The completed Georg von Neumayer Station is covered with snow for additional insulation. *In the foreground* are two red survival containers for use if a fire breaks out at the station. (Albrecht, Klaus-Peter (Private possession))

Fig. 11 A storage facility in an unheated tube at Georg von Neumayer Station. (Albrecht, Klaus-Peter (Private possession))

For the next ten years, the station would accommodate seven to nine people (one physician and station leader, two meteorologists, one or two geophysicists, one or two technicians, one radio operator, and one cook), until it became uninhabitable because of an increasing snow load well over ten meters in height.[41] The roofing ceremony was held on February 24, 1981, and the meteorological field station transmitted its data to the international network of weather services from February 25 onward.

To demonstrate a "German winter presence,"[42] a small crew (comprising technicians and one physician) was initially to overwinter at the station. However, as one technician had to be replaced, an Austrian student of meteorology, Friedrich Oberleitner (who had previously carried out meteorological and glaciological studies), was recruited on-site for the first scientific measurements.

The legal status of Georg von Neumayer Station was equated with that of a merchant ship; consequently, the leader of the station (i.e., the physician) had the responsibilities and privileges of a naval captain.[43] At that time, only radio contact with other nearby Antarctic stations was free of charge, while contact with home

---

[41] A list of the overwinterers from 1981 to 1991 is available in Reinke-Kunze, Christine: Aufbruch in die weiße Wildnis. Die Geschichte der deutschen Polarforschung, Hamburg 1992, pp. 409–411.

[42] Fleischmann: Zu den Kältepolen, p. 230.

[43] Puskeppeleit, Monika: Die wahren Abenteuer sind im Kopf, in: Sobiesiak, Monika/Susanne Korhammer (ed.): Neun Forscherinnen im ewigen Eis. Die erste Antarktisüberwinterung eines Frauenteams, Berlin 1994, pp. 129–130, 141.

Fig. 12 Five expedition helicopters brought the crew of the *Gotland II* to safety at the Lillie Marleen Hut on (approximately) December 22, 1981. (Kleinschmidt, Georg (Private possession))

was subject to charges. Once each week the crew sent a report to the AWI, which would forward it to various other institutions and persons. When Georg von Neumayer Station was completed on March 3, 1981, the FRG was acknowledged as the 14th equal member in the consultative party of the Antarctic Treaty states.[44] Now the FRG was entitled to take part in future votes or, if necessary, make use of its right of veto. On the occasion of the accession, the German Federal Post Office issued two commemorative stamps with face values of DM 1 and DM 1.10.

In the season of 1981–1982, the basic research activities of the BGR during GANOVEX II were to be continued with use of the freighter *Gotland II*.[45] First, the Lillie Marleen Hut was occupied. The expedition had hardly arrived in the Ross Sea when the *Gotland II* was surrounded by pack ice and sprang a leak. On December 18, 1981, the crew members at the Lillie Marleen Hut learned that the *Gotland II* had been damaged, and they listened on the radio as it sank the following night. Fortunately, all of the seamen and scientists could be rescued by the five expedition helicopters at the Lillie Marleen Hut (Fig. 12). Apart from the quantifiable damage, amounting to nearly DM 10 million, all previous scientific records, of inestimable value, were lost. This disaster was a bitter setback for the federal German Antarctic program. The geological fieldwork could be resumed only during GANOVEX III

---

[44] Kohnen: Antarktis-Expedition, p. 195.
[45] http://vphn-os.de/wp-content/uploads/2015/12/Auszug-aus-VN4-2015_Weihnachten-1981-Untergang-der-Gotland-II.pdf; Schmidt, Josef: Antarktisexpedition erleidet Schiffbruch. Deutsches Forschungsschiff im Packeis leckgeschlagen und gesunken, in: Süddeutsche Zeitung, issue of December 19/20, 1981.

(1982–1983) in the Lanterman Range (North Victoria Land) during the next southern summer.

According to a statement made by Heinz Kohnen (the leader of the expedition on the *Polarsirkel* as to the choice of location for the planned Georg von Neumayer Station) the management of the Antarctic Treaty member states in 1981 was essentially limited "to matters of research and environmental protection, but it is to be expected that the consultative round will also prepare a regime for the use of the resources and the continuation of the Antarctic Treaty in the decades lying ahead of us".[46]

Despite the scarcity of state funding at that time, the FRG invested "plenty of money for [its] 'admission ticket' to Antarctica. . . . [It] is a country with little natural resources and hence has a reasonable interest to partake in the resources of Antarctica."[47] Up until 1983, a total of approximately DM 300 million was made available for the Antarctic program. In the Antarctic research priority program of the German Research Foundation, a sum of 3.3 million euros per year was provided for the period from 2007 to 2010, which included the Fourth International Polar Year.[48] Afterward, the financial frame was reduced to below 3 million euros per year.

However, since the location on the Filchner–Ronne Ice Shelf was still of interest for research, Filchner Station was inaugurated on January 11, 1982, 1300 kilometers away from Georg von Neumayer Station (Fig. 13). Because of snowfall and snowdrift, it stood on stilts and had to be elevated by approximately one meter every two to three years.[49] It had space for 12 scientists and initially served as a starting point for glaciological and geophysical studies during summer within the scope of the Filchner–Ronne Ice Shelf Program. Here, the flow properties of the ice shelf, the substance load from the atmosphere into the ice, and the interactions between the ice shelf and the ocean were primarily studied.

Finally, the most modern polar research icebreaker in the world was christened with the name *Polarstern* on January 25, 1982, and departed on its first voyage to bring supplies to the German Antarctic station in the season of 1982–1983.[50] After the *Gauss*, the *Polarstern* was the second German polar research vessel, which, like *Gauss*, was built at the Howaldt Shipbuilding Yard in Kiel.

The *Polarstern* measures 118 meters in length and 25 meters in width.[51] Its molded hull depth amounts to 13.6 meters and its draft to 10.5 meters. The combined capacity of its four engines is 14,000 kilowatts (20,000 horsepower), and it

---

[46] Kohnen: Antarktis Expedition, p. 14.

[47] Mandelsloh/Freyenhagen: Antarktis, p. 64.

[48] Bundesministerium für Umwelt, Naturschutz und Reaktorsicherheit: Deutsches Engagement für den weißen Kontinent. Antarktisvertrag–30 Jahre Konsultativstatus, Berlin 2011, p. 33.

[49] Gravenhort, Gode: Die Antarktisexpedition 1981/82, in: Gode, Gravenhort: Antarktis-Expedition 1981/1982 (Unternehmen »Eiswarte«), in: Berichte zur Polarforschung 6/'82 (1982), pp. 9–17; Fleischmann: Zu den Kältepolen, p. 235.

[50] Mandelsloh/Freyenhagen: Antarktis, p. 80.

[51] Anonymous: Das Deutsche Polar-Forschungs- und Versorgungsschiff, in: Hansa 118 (1981) 17, pp. 1217–1220; Fleischmann: Zu den Kältepolen, pp. 238–242.

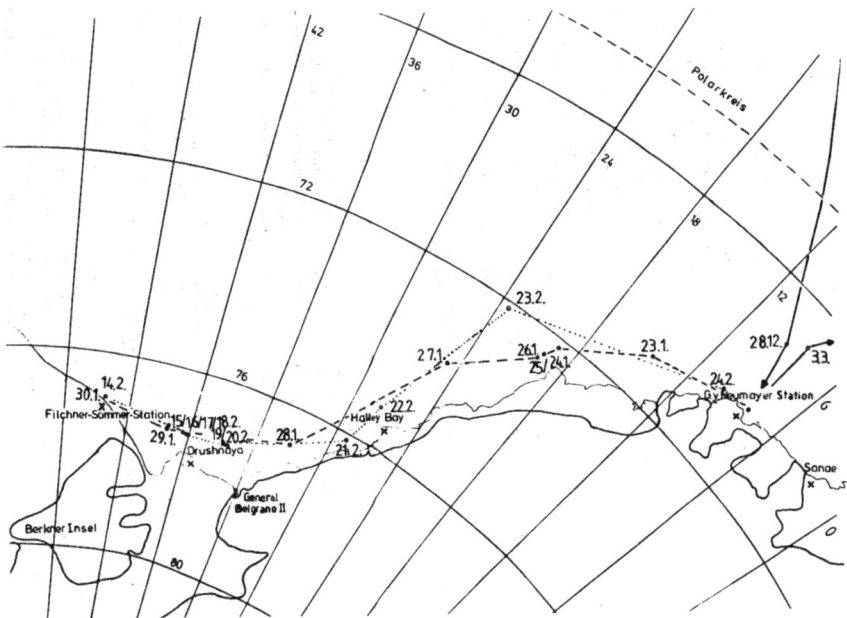

Fig. 13 The travel route of the *Polarqueen* in the eastern part of the Weddell Sea, between Georg von Neumayer Station and Filchner Station, early in 1982. (Gravenhort, Gode: Antarktis-Expedition 1981/1982 (Unternehmen »Eiswarte«), in: Berichte zur Polarforschung 6/'82 (1982), p. 17)

can reach a maximum speed of 16 knots. The ship can move at a constant speed through ice up to 1.5 meters thick and break up ice 2 meters thick by ramming it. An important planning feature was the installation of a 25-ton crane, with which heavy caterpillar tractors can be unloaded onto an ice shelf edge over 20 meters above the waterline. There is also a helicopter landing pad. The *Polarstern* requires a crew of 36 men and has room for 70 scientists and the overwintering crew. The official building costs amounted to DM 188 million (Fig. 14).

Independent of the AWI, the BGR set up the Gondwana Hut in North Victoria Land in 1983 to explore the geology of the Transantarctic Mountains from there.[52] In mid-January, Gotthilf Hempel, the founding director of AWI, proudly reported to *Süddeutsche Zeitung* that German polar research had "entered international business astonishingly fast" and said, "We are now involved in planning" and "[shaping] the future of the last free space on earth."[53] To provide further support for summer campaigns in the area surrounding Georg von Neumayer Station, two polar aircraft—a Do 128-6 (*Polar 1*) for transport missions and a somewhat larger Do 228-200 (*Polar 2*) from Dornier, reconstructed especially for Antarctica with a maximum transport capacity of 19 passengers and a facility for electromagnetic

---

[52] www.www.bgr.bund.de/DE/Themen/Polarforschung/Antarktis/Logistik/logistik_node.html

[53] Kistenmacher, Gert: Im Eiskeller ins Geschäft kommen, in: Süddeutsche Zeitung, issue of January 14, 1983.

Fig. 14 The *Polarstern* unloading supplies for Georg von Neumayer Station, on the edge of the Ekström Ice Shelf, on January 7, 1987. (Oerter, Hans (Private possession))

measurements—were purchased and successfully deployed for the first time in 1983–1984.[54]

Despite the high technological development of the Antarctic transport fleet, work in the south was still associated with dangers. On December 20, 1983, starting a skidoo (snowmobile) ended in a tragic accident, which was not survived by the geophysicist Klaus Wallner.[55] Later, the Wallner Summit in the Kottas Mountains [Kottasberge] (in the Heimefront Range, Neu-Schwabenland) was dedicated to him.[56]

Life at the permanent research stations is determined by the polar summer and the dark polar night. To structure outdoor work on days with 24-hour daylight or darkness, a very constant course of the day proves useful. The following example describes overwintering at Georg von Neumayer Station in 1984[57]:

8:00–9:00 a.m.: morning coffee
Midday: lunch
6:00 p.m.: supper
8:00 p.m.: meeting in the messroom for leisure-time activities

---

[54] Fleischmann: Zu den Kältepolen, pp. 248–257.

[55] Anonymous: Deutscher Physiker in der Antarktis tödlich verunglückt, in: Süddeutsche Zeitung, issue of December 23, 1983; personal communication from Ulrich Stuckenberg on January 15, 2015.

[56] www1.data.antarctica.gov.au/aadc/gaz/display_name.cfm?gazid=107149

[57] Herold, Werner: Alltag in der Eiswüste, in: GEO 1986 (3) 3, p. 44, 46, 50.

Fig. 15 The New Year's Eve banquet at Georg von Neumayer Station in 1984. As it is summer in the Southern Hemisphere, the overwinterers sitting around the meat fondue are a bit sunburned. Only later do they learn of the strong ultraviolet irradiation due to the ozone hole in the atmosphere. *From left to right*: Reinhard Beyer, Wolfgang Kobarf, Werner Herold, Joachim Schug, Heiko Muhle, and Detlef Knoop. (Schönhofer, Georg (Private possession))

The station leader Werner Herold realized very fast that whoever had a professional background as a seaman would cope more easily with this discipline than young scientists who reported for overwintering because of the allegedly absent constraints and would have preferred not to have a boss around. In particular, the winter routine work prevailed and the diversification of spare-time activities left much to be desired. The overwinterers became discontent and irritable. Only long conversations could prevent the group from breaking up (Fig. 15).

Selection of overwinterers at the West German Antarctic station was described by Gotthilf Hempel as follows: "We do not subject them to psychological tests, neither before [nor] during [nor] after overwintering, and we do not ask afterward how they got along."[58] Instead, a long interview was conducted with each participant, as was the case in the overwintering of 2015.[59]

In daily life, with its defined research program, a routine had established itself long ago. Daily radiosonde ascents and magnetic measurements were made, air samples were taken, and the occurrence of earthquakes was surveyed (Fig. 16).

---

[58] Sobiesiak/Korhammer: Neun Forscherinnen, p. 8.
[59] Kohlberg, Eberhard: Eberhard Kohlberg, der einen Koch für die Antarktis sucht, in: Süddeutsche Zeitung, issue of January 31/February 1, 2015.

Fig. 16 A daily launch of a radiosonde on a dark night during the southern winter of 1984. (Schug, Joachim (Private possession))

In the season of 1984–1985, the Federal Ministry of Research and Technology rented the *Polar 3* for four months to fly—together with the *Polar 2*, after a stopover at the American Amundsen–Scott Station at the South Pole—to the American McMurdo Station on the Ross Sea.[60] There, scientists from the BGR successfully conducted an electromagnetic survey of the region near Gondwana Station. The return flight went over the South Pole, over Punta Arenas (in Chile), and along the African Western Sahara coast. Although this flight route was internationally authorized, Polisario rebels mistook the research aircraft for Moroccan reconnaissance planes and shot down the *Polar 3* on February 24, 1985. The three-man crew—consisting of pilot Herbert Hampel, copilot Richard Möbius, and mechanic Josef Schmidt—did not survive the crash. Because loss of the airplane as a result of warfare was not covered by insurance, the federal minister of research and technology had to provide new funding to commission the *Polar 4* as a further development of the Do 228 (Fig. 17).

In the mid-1980s, it was discovered that the ozone concentration over the South Pole was declining and the so-called ozone hole was extending beyond Antarctica, thereby decreasing the protection of southern continental regions against harmful ultraviolet radiation.[61] Research on the causes resulted in new research projects in Antarctica.

---

[60] Fleischmann: Zu den Kältepolen, p. 252.

[61] Anonymous: Loch in der Ozonschicht wird größer, in: Süddeutsche Zeitung, issue of June 30, 1986.

Fig. 17 The polar research aircraft *Polar 4* in the outdoor section of the Dornier Museum in Friedrichshafen, photographed in 2013. (Lüdecke, Cornelia (Private possession))

At the beginning of the southern summer season in October 1986, Drescher Station was built on the Riiser-Larsen Ice Shelf, on the northeastern coast of the Weddell Sea, as a mobile ice camp for four to five persons to study the nearby colony of emperor penguins and the assemblage of Weddell seals during the summer.[62]

In January 1987, a drill camp was established in the Ritscher Upland, nearly 220 kilometeres south of Georg von Neumayer Station (Fig. 18). The *Polar 4* primarily served the purpose of connecting the drill camp and the station, and flying the ice cores back.[63] During GANOVEX V in the season of 1988–1989, the BGR installed a large field camp in central Victoria Land (Fig. 19).[64] Rock samples collected in the surroundings were flown out by helicopter (Fig. 20).

The hopes nurtured by many parties to extract ore or valuable minerals by industrial mining operations in Antarctica were dashed, as the mineral resources were covered by a 2000-meter-thick layer of ice.[65] The Antarctic Information and Protective Association [Antarktische Informations- und Schutzgemeinschaft], headquartered in Karlsruhe, was already issuing warnings about environmental pollution at polar stations and especially about the unforeseeable risks of oil drilling in shallow water areas. Like Greenpeace and other environmental protection

---

[62] www.awi.de/de/infrastruktur/stationen/drescher_station

[63] Miller, Heinz/Oerter, Hans (ed.): Die Ekström-Traverse 1987, in: Berichte zur Polarforschung 57/90 (1990), pp. 53–55.

[64] Personal communication from Georg Kleinschmidt on January 28, 2015.

[65] Jung-Hüttl, Angelika: Kaum Chancen für Antarktis-Schätze, in: Süddeutsche Zeitung, issue of January 19, 1987.

Fig. 18 The BC-1 ice-coring camp in the Ritscher Upland, photographed in January 1987. The automatic weather station is seen *in the foreground*, with two accommodation containers *in the background*, and, *in between them*, the *Polar 4*. (Oerter, Hans (Private possession))

organizations, it demanded the establishment of a "World Park Antarctica."[66] Nevertheless, after exploitation of raw material deposits was regulated in Wellington (New Zealand) on June 2, 1988, word of a "Race to Antarctica" began to spread.[67]

Enforcement of this regulation failed because of the Convention on the Regulation of Antarctic Mineral Resource Activities (CRAMRA).[68] At that time, the question was raised as to whether the countries conducting research in Antarctica would behave like a multinational model family and abide by all regulations. The answer given by Gotthilf Hempel was not very optimistic; he doubted they would do so "[while] the scientific interest is surpassed by the economic."[69] Finally, in a resolution passed in 1989, the United Nations demanded that Antarctica be granted the status of a world nature reserve.[70]

---

[66] Antarktische Informations- und Schutzgemeinschaft: Antarktis—das letzte Paradies? Koblenz 1986, pp. 106–107, 113–117.

[67] Anonymous: Wettlauf zur Antarktis: Präsent sein ist alles, in: Der Spiegel (1988) 27, pp. 132–134.

[68] Headland: Chronology, p. 522.

[69] Anonymous: Wettlauf zur Antarktis, p. 134.

[70] www.umweltbundesamt.de/sites/default/files/medien/pdfs/usp.pdf

Antarctic Research in the Federal Republic of Germany Since 1975 215

Fig. 19 Freeze-dried laundry at the field camp in the Kavrayskiy Hills in Oates Land (central Victoria Land), photographed at the end of January 1989. (Kleinschmidt, Georg (Private possession))

Fig. 20 Loading of samples at Camp Capsize in central Victoria Land on December 28, 1988. (Kleinschmidt, Georg (Private possession))

The Merging of West and East German Antarctic Research After 1990

The first Antarctic overwintering of an all-female team took place at Georg von Neumayer Station in 1990 (Figs. 21 and 22).[71] During the winter, the sunspot cycle reached its maximum, which not only was associated with an abundance of fantastic polar lights but also interfered with radio communications.[72]

Radio contact between FRG women and GDR men at Georg Foster Station was initially not considered desirable by the GDR for political reasons, but after the political change, these rules no longer applied. Now both groups initiated regular radio communication contact. The news about the opening of the borders between the FRG and the GDR was received with disbelief by the crew members at Georg Forster Station (Fig. 23).[73] What was in store for them in Antarctica? Who would come and collect them now that the national structures were on the brink of collapse? What changes would there be at home, and what would it be like when they

Fig. 21 The globally unique all-female overwintering team in 1989–1991. *From left to right, front row*: Susanne Baumgart, Susanne Korhammer, Monika Sobiesak, and Ursula Weigel; *back row*: Grazyna Muhle, Ulrike Wyputta, Monika Puskeppeleit, Estella Weigelt, and Elisabeth Schlosser. (Puskeppeleit, Monika (Private possession))

---

[71] Sobiesiak/Korhammer: Neun Forscherinnen.

[72] Baumert, Susanne: Zu Wasser, im Eis und durch den Äther, in: Sobiesiak/Korhammer: Neun Forscherinnen, pp. 117–120; Puskeppeleit: Die wahren Abenteuer, pp. 143–144.

[73] Fleischmann: Zu den Kältepolen, pp. 284–285.

The Merging of West and East German Antarctic Research After 1990 217

Fig. 22 First aid instruction in the messroom at Georg von Neumayer Station. *From left to right*: Muhle, Weigel (*standing*), Schlosser, Weigelt, and Puskeppeleit. (Puskeppeleit, Monika (Private possession))

Fig. 23 Georg Forster Station, looking northward, with the radio complex *in the right foreground*, photographed in 1991. (Strecke, Volker (Private possession))

returned? Both groups learned about the events only by radio or teleprinter, without any knowledge of the corresponding television images and newspaper information. Apart from this, in their Sunday talks, they exchanged not only their magnetic measurement data but also the experiences of the station leaders and—very importantly for overwinterers—new cooking recipes. When West German communications with the outside world were very restricted because a transmitter had broken down, the GDR station served as a relay station for official and private mail. Finally, after all of the routine work was done, the West German *Polarstern* collected overwinterers from the GDR station in the southern autumn of 1990 and delivered a replacement team.

At home, consultations about the future of GDR polar research had already begun. The GDR was represented by Hans-Jürgen Paech, who had been the coordinator of Antarctic research at ZIPE since May 1989 and was committed to upholding polar research in his country. The interests of the FRG were represented by Gotthilf Hempel, the director of the AWI, which ultimately took on the leadership of all German polar research. Since the AWI was not willing to maintain two polar stations and the planning of Neumayer Station II was already under way, while maintenance of Georg Forster Station according to environmantal protection was questionable, the GDR station was to be completely dismantled.[74] Before that, however, routine measurements continued and a joint polar light research program in association with the Meteorological Institute in Helsinki (Finland) was finalized.

On October 3, 1990, when the German Reunification Treaty was signed and the GDR no longer existed after its accession to the FRG, many congratulatory telegrams arrived at Georg von Neumayer Station and there was long radio contact with colleagues from the neighboring Antarctic stations. On this festive occasion, the East German flag was replaced by the West German flag at Georg Forster Station, but everything else remained unchanged.

As a matter of principle, the West German women's team at Georg von Neumayer Station strove to achieve an environmentally conscious overwintering and succeeded (for the first time ever) in getting the AWI's approval for complete waste separation at the station, composting of organic waste, and return of all refuse to Germany.[75] This put an end to the hitherto common combustion of waste. In addition, they introduced wind energy from two 1-kilowatt wind power turbines as an alternative energy resource for the station. Actually, the women's team hardly differed from the men's teams. Snow had to be shoveled in all kinds of weather to refill the snow-melting unit so the station would always have enough water. As storms with strong snowdrifts occur all the time, the shovel remains the most important tool for overwintering (Figs. 24, 25 and 26).

---

[74] Hempel, Gotthilf: Deutsche Beiträge zur europäischen Zusammenarbeit in der Polarforschung, in: Polarforschung, 60 (1990) 3, p. 246; Fleischmann: Zu den Kältepolen, pp. 284–289.

[75] Muhle, Grazyna/Korhammer, Susanne: Vom Dieselgenerator bis zur Spülmaschine, in: Sobiesiak, Monika/Susanne Korhammer (ed.): Neun Forscherinnen im ewigen Eis. Die erste Antarktisüberwinterung eines Frauenteams, Basel, Boston, Berlin 1994, pp. 66–70; see also Jung-Hüttl, Angelika: Forschen im Kühlschrank der Erde, in: Kosmos (1992) 11, p. 59.

Fig. 24 Ground drift at the survival containers at Georg von Neumayer Station, photographed in 1990. (Puskeppeleit, Monika (Private possession))

Fig. 25 Refilling the snow-melting facility at Georg von Neumayer Station, under strong snow-drift conditions and at an outdoor temperature of −35 °C, in 1990. (Puskeppeleit, Monika (Private possession))

Fig. 26 Shoveling a Scott tent free of snow after a snowstorm. (Puskeppeleit, Monika (Private possession))

After their successful overwintering, they were congratulated by Heinz Kohnen, the head of logistics at the AWI: "You definitely have contributed to overcoming prejudices. The whole Antarctic scenery has followed your work with attention. As we say in polar research circles: you have broken the ice."[76] In fact, they had broken the ground for mixed overwintering teams at German polar stations. However, this first and globally unique all-female overwintering in the history of Antarctica received more recognition internationally than it did at home.[77]

Independent of the AWI, the German Aerospace Center [Deutsches Zentrum für Luft- und Raumfahrt (DLR)] opened the German Antarctic Receiving Station (GARS) at O'Higgins on the northern point of the Antarctic Peninsula near the Chilean General Bernardo O'Higgins Station. The station served the purpose of receiving measured data from the satellite European Remote Sensing (ERS)-1, which had just been launched for earth observation, continental drift surveys, and other surveying tasks.[78]

---

[76] Puskeppeleit: Die wahren Abenteuer, p. 145.

[77] Edwards, Kerry/Graham, Robyn (ed.): Gender on ice, in: Proceedings of a conference on women in Antarctica held in Hobart, Tasmania, under the auspices of the Australian Antarctic Foundation, 19–21 August 1993, Canberra 1994; Martin, Stephen: A History of Antarctica, 2nd edition, New South Wales 2013; see also Fleischmann: Zu den Kältepolen, pp. 232–233; Puskeppeleit, Monika: The allfemale expedition: Personal perspective, in: Edwards/Graham: Gender on ice, in, pp. 75–81; Puskeppeleit, Monika: The untold story: The German allfemale overwintering, in: Edwards/Graham: Gender on ice, in, pp. 49–52.

[78] https://www.dlr.de/eoc/en/desktopdefault.aspx/tabid-9472/16238_read-40703/

When the environment protection protocol for Antarctica was ratified as an amendment to the Antarctic Treaty in Madrid on October 4, 1991, one article prohibited all activities related to mineral resources (such as mining) and allowed only purely scientific exploration.[79] Furthermore, all activities that could have an impact on the environment or ecosystems were strictly regulated. This environment protection protocol can be renegotiated only after 50 years.

Another event, in March 1991, was the arrival of the new overwinterers at Georg Forster Station—the "ultimately last really existing GDR biotope" since October 1990, despite Germany's reunification; after all, the entire preparation for overwintering had taken place during GDR times and nothing had changed at Georg Forster Station since the reunification.[80] "Socialism sent us to Antarctica to live [for] one year in communism with the Russians, and in the end we will return to a capitalist home country,"[81] said one of the old overwinterers with deep concern, for they had heard on the news that their occupational perspectives would be absolutely uncertain after returning home. Only four overwinterers wanted to take part in the fourth Antarctic expedition (1991–1993), which had been prepared for by the GDR. During the southern summer of 1991–1992, when 13 "new ones" were flown in by plane, Wieland Adler and the station leader, Gerold Noack, climbed, for the first time, from their base camp at Lake Untersee up to the 3050-meter-high Ritscher Summit on December 17, 1991.[82] On its way back to Germany, the *Polarstern* took the first freight from the station homeward in March 1992.

The reconstruction of polar research in the reunited Germany came to an end when the Research Office at the AWI was inaugurated on Potsdam's Telegrafenberg on March 11, 1992. It was supposed to promote cooperation between the polar researchers of the former GDR and the FRG, and uphold contacts with Eastern Europe and Russia.[83] In particular, it was intended to "maintain the experiential potential for German polar research which had grown under difficult circumstances."[84]

This merger complemented—in a profitable way—West German Antarctic research, which, apart from its routine measurements, was geared more toward maritime and biological issues than East Germany's terrestrial Antarctic research.

Concomitantly, Georg Forster Station was closed forever, according to plan, in February 1993, after a service life of 17 years.[85]

---

[79] www.umweltbundesamt.de/sites/default/files/medien/pdfs/usp.pdf

[80] Noack, Gerold: Antarktis. Abenteuer Wissenschaft—Ein Lausitzer im Ewigen Eis, Cottbus, pp. 51–54, 351.

[81] Ibidem, pp. 53–54.

[82] Ibidem, pp. 278–299, 354.

[83] Hempel, Gotthilf: Blühende Landschaften im Ewigen Eis, in: Polarforschung 79 (2009) 3, p. 190; Hempel: Deutsche Beiträge, p. 246; Fleischmann: Zu den Kältepolen, pp. 284–289.

[84] Fleischmann: Zu den Kältepolen, p. 285.

[85] Berichte zur Polarforschung 1994, No. 152/94, pp. 219–220; Lange: Sonne, Sturm und weiße Finsternis, pp. 256–261.

Fig. 27 The tube complex at Neumayer Station II, photographed from a helicopter in 1992. (Strecke, Volker (Private possession))

Important measurements such as ozone soundings were then taken over by Neumayer Station II, which had opened on March 31, 1992 (Fig. 27).[86] Today, the unbroken continuation of measurements constitutes the longest recording period of vertical ozone profiles over Antarctica.

In accordance with the environmental protection protocol, Georg Forster Station was completely removed and disposed of prior to 1996.[87] Only a bronze plaque commemorates the former GDR station (Fig. 28). In 2013, the location of Forster Station became the 87th entry in the list of Historic Sites and Monuments in Antarctica.[88]

Apart from this, the AWI received funding to open the Dallmann Laboratory, maintained jointly with the Instituto Antárctico Argentino, in January 1994. It consists of four laboratory containers set up near the Argentinian Carlini Station (formerly known as Jubany Station) on King George Island, one of the South Shetland Islands north of the Antarctic Peninsula.[89] The laboratory enables cooperation between Argentinian, German, and Dutch scientists in the southern summer from October to March.

---

[86] www.awi.de/de/infrastruktur/stationen/neumayer_station; Fleischmann: Zu den Kältepolen, p. 233
[87] Lange: Sonne, Sturm und weiße Finsternis, pp. 262–271.
[88] www.ats.aq/documents/ATCM36/WW/atcm36_w w004_e.pdf
[89] www.awi.de/de/infrastruktur/stationen/dallmann_labor

Fig. 28 A bronze plate commemorating the 20-year existence of the German Democratic Republic's Georg Forster Station. (Gernandt, Hartwig (Private possession))

German Antarctic research suffered problems from a gigantic plate breaking off from the Filchner–Ronne Ice Shelf, on which Filchner Station used to stand (Fig. 29).[90] This was discovered by British researchers on satellite images taken on October 13, 1998. For further tracking purposes, the 150-kilometer-long, nearly 35-kilometer-wide, and 200- to 400-meter-thick ice island was referred to as A-38. Since the summer season in the Antarctic had not yet begun, Filchner Station was fortunately not occupied, but everything of value had to be salvaged immediately.

Another problem was that the Antarctic Treaty demanded that in the event of decommissioning, a station had to be dismantled and completely disposed of. Hence, in the season of 1998–1999, apart from having to bring supplies to Neumayer Station II, the *Polarstern* was assigned to salvage Filchner Station. After it had moored at the ice edge of A-38 on January 31, 1999, the fuel barrels were the first things collected from the station, which was 25 kilometers away from the ice edge. With the exception of the steel framework of the station's platform, everything was loaded on board the *Polarstern* within 10 days (Figs. 30 and 31). With the loss of Filchner Station, research on the Filchner–Ronne Ice Shelf had come to a premature end. The death of the cook at Neumayer Station II in April 1999 darkened the annual balance of 1998–1999 even more.[91]

Within the scope of the European Project for Ice Coring in Antarctica (EPICA), which undertook to explore climate history dating back 500,000 years, a

---

[90] www.antarktisstation.de//eisberge; https://antarktis.ch/2000/01/01/antarktis-xvi-1998-99/; https://de.wikipedia.org/wiki/Filchner-Station; Fleischmann: Zu den Kältepolen, pp. 235–236.
[91] https://antarktis.ch/2000/01/01/antarktis-xvi-1998-99/

Fig. 29 Filchner Station on the broken-off ice shelf plate, photographed in January 1999. (Fahrbach, Karin (Private possession))

Fig. 30 Filchner Station dismantled down to its foundation platform. (Fahrbach, Karin (Private possession))

preliminary exploration had been conducted in the season of 1995–1996, and accommodation containers were brought from Neumayer Station II to the base camp (Kottas Camp) in the Kottas mountains, using tracked PistenBully vehicles

Fig. 31 Parts of the dismantled Filchner Station brought to the edge of the ice with PistenBully vehicles. (Fahrbach, Karin (Private possession))

(Figs. 32 and 33).[92] From January 20 to January 24, the vehicles got stuck in a heavy snowdrift before the crew members could free themselves again (Fig. 34). On the way back in February, a snowstorm with a so-called whiteout hindered the drive, as the diffuse illumination meant that contrasts were no longer discernible.

In 1988, a snow pit was dug out in Dronning Maud Land, which clearly displayed layers of annually added snow. Snow samples were taken from various depths to determine the snow's density, which correlates with the water content of the sample and contains information about annual precipitation rates when allocated to the annual layers (Fig. 35).

In the summer of 2000–2001, Kohnen Station was finally established for the second deep-ice-coring operation of EPICA, 750 kilometers away from Neumayer Station II, at an altitude of 2892 meters on the ice cap, to learn more about the earth's climate in the past by analyzing ice cores.[93] The station consisted of eleven 20-foot containers, which were put together on a platform nearly two meters above the snow surface (Fig. 36). The platform of the station was elevated every two years by about 60 centimeters to make up for added snow. The containers housed a

---

[92] Personal communication from Hans Oerter, Rotenburg, on January 22, 2015.

[93] Oerter, Hans/Cord Drücker/Sepp Kipfstuhl/Frank Wilhelms: Kohnen Station—the Drilling Camp for the EPICA Deep Ice Core in Dronning Maud Land, in: Polarforschung 78 (2008), pp. 1–23; www.awi.de/de/infrastruktur/stationen/kohnen_station. The station was named after the geophysicist Heinz Kohnen, who was head of the logistics department at the Alfred Wegener Institute from 1982 to 1997.

Fig. 32 An aerial view of Kottas Camp, with the Kottas Mountains *in the background*, during the southern summer of 1996–1997. *In the left foreground* are a tank container, a freight sledge with a 10-foot container, and a generator. *To the right* there is a kitchen tent; *behind it, on the left*, is an accommodation container. *In the background*, two large "tomatoes" (survival containers) and three Scott tents for accommodation are visible. *To the right of the kitchen tent* stands the lone toilet outhouse. *On the right side, in serial order from bottom to top*, are two freight sledges, one freight container, one tank container, and three PistenBully vehicles. (Oerter, Hans (Private possession))

generator, a snow-melting facility, a workshop, a warehouse, sleeping quarters, a toilet and washroom, a kitchen, a messroom, and the important telecommunications. Next to the station, there were further sleep modules, food storage containers, and tank containers, so that a total of 20–25 people could work there.

In early 2001, a trench measuring 66 meters in length, 4.6 meters in width, and 6 meters in depth was dug out for the coring work and covered with a wooden roof. Here, the drill rig and spaces for intermediate storage and processing of ice cores were installed. The coring work began in January 2001 and ended in January 2006 (Fig. 37).

An insight into the everyday routine of a mixed crew at Neumayer Station II has been given by Nora Graser.[94] On November 24, 2006, all overwinterers met in Frankfurt in order to fly together to Cape Town, where they arrived 11 hours later. As the weather conditions were initially unfavorable for a flight on an Ilyushin IL-76 to the ice airport near the Russian Novolazarevskaya Station at the Schirmacher Oasis, the stopover in South Africa lasted five days instead of two. After two more overnight stays at the Russian station because of the weather, the crew members flew

---

[94] Graser, Nora: Kalte Füße inklusive. Mein Jahr in der Antarktis, München 2008.

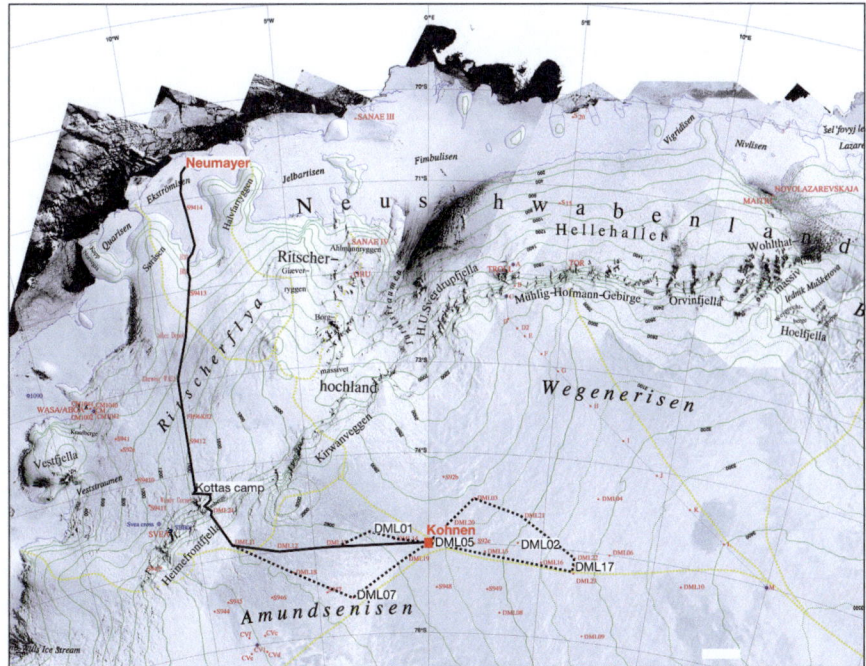

Fig. 33 A map of Dronning Maud Land, showing the route of the Kottas traverse (the route to Kottas Camp), as well as other routes and the location of Kohnen Station. (Oerter, Hans (Private possession))

Fig. 34 Snow shoveling after a sojourn of several days en route to Kottas Camp through a strong snowdrift in January 1996. (Oerter, Hans (Private possession))

Fig. 35 A snow pit at point DML21 in Dronning Maud Land, photographed on January 4, 1998. To determine the density of the snow, 500-milliliter snow samples are punched out with coring sleeves and weighed. (Oerter, Hans (Private possession))

on a Basler BT-67 (an aircraft especially equipped for deployment in polar regions) to the German Neumayer Station II, which they reached within three hours, nine days after their takeoff from Frankfurt. This was a very fast arrival, compared with the long sea voyages during the first two Antarctic expeditions about 100 years ago. In Antarctica, the overwinterers from the previous year were awaiting the new arrivals, whom they would familiarize with the station work in the coming months.

To accommodate the "newcomers," every chamber had to be shared by two people. Orientation inside the 90-meter-long tubes, which were now under a 12-meter-thick layer of snow, proceeded according to the cardinal points: the western tower, the eastern tube, and the northern ramp. To get outside for daily routine walks over the towers to the meteorological or geophysical observatories, one had to climb up 90 grating steps with one's equipment. In January, while the supply work was still in progress, one member of the documentary film team was run over by a PistenBully

Fig. 36 Kohnen Station, photographed during the European Project for Ice Coring in Antarctica (EPICA) in 2006. (Oerter, Hans (Private possession))

and seriously injured. Two days later he was flown by helicopter to the Norwegian Troll Station in Antarctica, where a chartered private jet evacuated him to Cape Town. He was treated at a hospital there and, after two weeks, was out of life-threatening danger.

Once the new arrivals had become familiar with all facilities and their operation, they officially took over the responsibility for the station. A written protocol sealed this solemn act in mid-January 2007. The new group was now left behind alone. However, in the meantime, communications technology had become so highly advanced that the overwinterers had internet access and could e-mail or chat and phone home at all times.

Despite the comparably extensive comfort at the station, serious accidents still happened in Antarctica. Early in March 2008, a helicopter crashed on its way to the station and caused the sudden and violent deaths of the pilot, Stefan Winter, and a Dutch scientist, Willem Poland, while three other passengers were seriously injured and had to be flown by the second helicopter to the *Polarstern* for further treatment.[95]

One year later, on February 20, 2009, 6.5 kilometers away from the old station and 16 kilometers from the ice shelf coast, Annette Schavan, Germany's federal minister of education and research at the time, inaugurated Neumayer Station III in

---

[95] www.awi.de/de/infrastruktur/schiffe/polarstern/wochenberichte/alle_expeditionen/ant_xxiv/ant_xxiv3/9_maerz_2008

Fig. 37 Drilling operations at Kohnen Station in 2006. The drill inside the tube of the 12-meter-long, swivel-mounted drill tower (the lower 6 meters are inside the drill hole) is being eased off into the drill hole. (Oerter, Hans (Private possession))

a festive ceremony via a live satellite connection all the way from the 13,758-kilometer-distant city of Berlin.[96] Neumayer III would be a laboratory with hitherto unknown possibilities, with which a new era would begin for the scientists. The station, which had cost 40 million euros to build, consists of 100 containers under a protective cover, standing on pillars (Fig. 38). This construction permits a hydraulic system to annually adjust the platform of the station (including its living quarters and work spaces) to the snow surface, which rises by approximately 90 centimeters each year. In addition to this platform above the ice surface, there is a huge garage under the snow cover, with workshops and storage facilities, which is accessible by an extensive lockable ramp. A 30-kilowatt wind power plant adds to the regular power supply of the station, which uses diesel generators (Fig. 39).

The current annual costs for the overwinterers and the maintenance staff in the summer season, as well as the material and transport costs, amount to 6–7 million euros, which comes from the basic financing of the AWI. The service life of Neumayer Station III is estimated to be at least a further 25–30 years.

---

[96] AWI press release from February 20, 2009; www.awi.de/de/infrastruktur/stationen/neumayer_station

Fig. 38 New food rations being supplied on December 20, 2011, for overwintering at Neumayer Station III. (Alfred-Wegener-Institut / Thomas Steuer, licensed under the terms of CC-BY 4.0 (https://creativecommons.org/licenses/by/4.0/legalcode))

Fig. 39 The north side of Neumayer Station III, with the open ramp and the wind power plant *in the foreground*, photographed in mid-December 2011. (Alfred-Wegener-Institut / Thomas Steuer, licensed under the terms of CC-BY 4.0 (https://creativecommons.org/licenses/by/4.0/legalcode))

Fig. 40 An ozonesonde is released from the meteorological observatory on the roof of Neumayer Station III on December 17, 2011. ((detail) Alfred-Wegener-Institut / Thomas Steuer, licensed under the terms of CC-BY 4.0 (https://creativecommons.org/licenses/by/4.0/legalcode))

Fig. 41 The air chemistry observatory, situated away from Neumayer Station III, photographed in mid-December 2011. (Alfred-Wegener-Institut / Thomas Steuer, licensed under the terms of CC-BY 4.0 (https://creativecommons.org/licenses/by/4.0/legalcode))

Fig. 42 The research aircraft *Polar 6* being prepared for a flight near Neumayer Station III on December 17, 2011. (Alfred-Wegener-Institut / Thomas Steuer, licensed under the terms of CC-BY 4.0 (https://creativecommons.org/licenses/by/4.0/legalcode))

Some facilities are now situated comfortably on the roof of the station, so that no one has to leave the station anymore to release radiosondes or ozonesondes (Fig. 40). The air chemistry observatory is placed away from the station to permit undisturbed measurement of atmospheric trace gases, given the extreme air purity at this isolated location in Antarctica (Fig. 41). These circumstances makes the measuring station one of the most important reference stations for clean air.

In addition to the previous main research topics, which, apart from trace gas analysis, also include studies on environmental changes, the sounds of seals and whales and noise from colliding icebergs have been recorded near the ice boundary with a hydrophone below the ice shelf and transmitted by a direct satellite connection to the AWI. These recordings are uploaded as a live stream to the internet under the name of PALAOA (Perennial Acoustic Observatory in the Antarctic Ocean).[97] In this way, the impacts on animals resulting from ship engine noises or experiments working with sound propagation (e.g., echo sounding) are to be documented for a certain period of time.

New perspectives on exploring the cryosphere (i.e., the ice regions on earth) were enabled by the successful launch of the CryoSat-2 satellite on April 8, 2010.[98]

---

[97] These sound measurements were recorded and could be accessed on the AWI homepage.

[98] www.esa.int/Our_Activities/Observing_the_Earth/CryoSat/Introducing _CryoSat; www.esa.int/ger/ESA_in_your_country/Germany/Drei_Jahre_CryoSat_Das_neue_Bild_vom_irdischen_Eis

It is operated by the European Space Agency (ESA) and involves the participation of several German institutes. Because it has a special radar sensor, it can precisely measure not only the thickness of the ice but also the height of the sea level. This makes it possible to determine alterations at the boundaries of the ice shelves, large ice fields, and glacier tongues in the polar oceans.

By 2015, the new aircraft *Polar 5* and *Polar 6* were in use and a new icebreaking supply vessel was in a state of planning to replace the old *Polarstern* prospectively in 2019. In 31 years the *Polarstern* has covered more than 1.5 million nautical miles (more than 2.7 million kilometers) (Fig. 42).[99] However, according to the timetable for 2020, the *Polarstern* still is in use.

German polar research—which, after its interruption by World War II, was successfully continued at first by the GDR and then also by the FRG—has contributed to Germany becoming one of the most important member states in the Antarctic Treaty.[100]

---

[99] www.awi.de, press release from July 22, 2014.

[100] Headland: Chronology, p. 32; see https://www.ats.aq/devAS/Parties?lang=e, see also Krause, Reinhard A./Salewski, Christian, R.: Das Alfred-Wegener-Institut Helmholtz-Zentrum für Polar- und Meeresforschung. Chronik und wissenschaftshistorische Grundlagen. Bremerhaven (2013/14).

The Future of German Antarctic Research

The number of countries with a consultative status in the Antarctic Treaty has grown from 12 in 1961 to 29 in 2009. It is not very useful to carry out only national programs at the stations instead of exploring Antarctica together and in a coordinated way. The Antarctic Treaty is increasingly coming under pressure because of the necessity to protect the environment on one hand and ward off economic interests on the other. The established marine protection areas, regulation of bioprospection, high penalties for environmental damage, and international rules that apply to Antarctic tourism must now be united by comprehensive environmental management.

The icy continent is currently subject to strong changes: gigantic floes are breaking off from the ice shelves, glaciers on the Antarctic Peninsula are retreating, the ocean circulation is changing, and the ozone hole is decreasing. These processes have a worldwide impact on the climate, on the height of the sea level, on biodiversity, and, ultimately, on humanity. The number of Antarctic tourists is increasing. About 44,600 people made a landing on the Antarctic Peninsula or the continent during the season of 2018–2019, indicating that the risk of introducing foreign species into this part of "untouched" nature is growing (Fig. 1). In addition, poorly equipped ships could easily become the cause of oil catastrophes, which are extremely difficult to clean up in the cold south. For example, powerful national interests in Antarctica have resulted in the so-called ABC problem—the overlapping (frozen) territorial claims of Argentina, Great Britain, and Chile. All countries committed to Antarctica (including Germany) maintain their permanent presence there by renewing old stations and building new ice-going ships for their maintenance. As some countries cannot raise the necessary funds to gain access to Antarctica on their own, they use the logistics of other nations or run joint stations that work according to most modern standards, such as the Belgian Princess Elisabeth Antarctica Station in Queen Maud Land, which has been supported by the International Polar Foundation since its opening on February 15, 2009.

Nowadays, scientists at the South Pole study the earth's climate, explore the lakes and mountain ranges under the icecap—as well as the deep sea surrounding

Fig. 1 Zuckerhut ["Sugarloaf"] (*left*) with a foehn wall in South Georgia, photographed on December 10, 1911. (Joester, Erich (Private possession))

Antarctica—and, on the basis of astronomical observations, contemplate the origin of life and the universe.

In April 2014, the Scientific Committee on Antarctic Research (SCAR) conducted the first Antarctic and Southern Ocean Science Horizon Scan in Queenstown (New Zealand) to determine the central problems of future Antarctic research. To this end, 75 scientists, attorneys, and political decision-makers from 22 countries (including five German experts) met to filter out the 80 most relevant of the 866 scientific research issues previously compiled by Antarctic researchers (Fig. 2).

Among the most important subjects are exploration of the range of the Antarctic atmosphere and its impact on the earth's energy balance, and the temperature gradient between the north and the south, as well as general air circulation and air chemistry. How the atmosphere interacts with the ocean and the ice coverage, and thus influences climate change, is as little known as the potential changes in local and global air circulation and the climate as a result of the current recovery from ozone depletion over the South Pole or the rise in greenhouse gases due to the thawing of permafrost. In addition, the oceans absorb heat and are becoming increasingly acidified as a result of carbon dioxide dissolving in seawater, which induces additional feedback effects that are not yet understood, including those affecting the deep sea. Nor do we know what determines seasonal change—i.e., the spread and volume of the sea ice cover, which, in winter, extends from the Antarctic coast for up to 1000 kilometers northward. Although the Antarctic icecap has remained stable for

Fig. 2 Scientists and organizers involved in the first Antarctic and Southern Ocean Science Horizon Scan in Queenstown (New Zealand) on April 22, 2014, after compiling the most significant tasks in Antarctic research for the next 20 years. (Kennicutt, Chuck (Private possession))

several thousand years, it is now appearing to lose considerable amounts of ice in several places for unknown reasons. Perhaps a threshold limit for the concentration of carbon dioxide has been exceeded and has triggered this effect. Or have warm water masses changed the ice shelf from below? New studies will provide answers.

As far as the geology of Antarctica is concerned, we mostly know of rock samples from the coastal region, whereas nothing is known about what the structure of the earth's crust underneath the icecap looks like, how volcanism and heat from the interior of the earth take effect, or how the continent of Gondwana broke apart. In this context, geological information referring to previous sea level heights are of interest, as they explain the comings and goings of interglacials and glacials, and the climate changes associated with them. The adaptation, diversification, and extinction of animals and plants are also connected with strongly varying environmental conditions. Is the Antarctic evolution rate unique, does an irreversible threshold limit exist, and which species is the first to react to these changes? (Fig. 3).

Now that the nearly 3000-meter-high icecap of Antarctica is extremely dry and cold, it is perfectly suited as a site for astronomical observations. A subglacial field of sensors for the international high-energy IceCube Neutrino Observatory, which has been operated by several German institutions near the South Pole since 2010, is capturing neutrinos in a cube with an edge length of one kilometer to learn something about sources of cosmic radiation. The lakes under the ice cover resemble the

Fig. 3 View from the Kottas Mountains onto the ice plateau of Antarctica, photographed on February 16, 2010. (Oerter, Hans (Private possession))

conditions on Jupiter and the icy moons of Saturn, and can therefore exemplarily serve as models. Furthermore, collections of meteorites can be used to decipher the origin of the solar system. A research topic of particular importance is large eruptions on the surface of the sun, called solar flares, which are capable of interrupting the global communication network.

Other than that, the impact of human activities on Antarctica must be observed because it is uncertain whether the previous provisions are adequate to protect the Antarctic environment and preserve its intact ecosystem. Are the previous regulations sufficient to control access to Antarctica?

To answer these questions, continuous financial support must be made available so that long-lasting meteorological and other geophysical measurement series are not interrupted, as has previously been the case when the oil price was increased and when the USA faced economic problems in 2013. Measurement programs that run for several decades appear not to be well secured financially in the long term, given the currently common support periods of three to five years.

In addition, access to the whole of Antarctica, and especially to the oceans surrounding it, must be made possible all year round, and environmental protection must be strengthened. In this regard, new technologies such as remote-controlled drones, vehicles, submersibles, and automatic measurement stations should be developed so that in the future, scientists will leave their footprints only during the summer months. New sensors for circumpolar satellites such as ICESat (Ice, Cloud, and Land Elevation Satellite), with which the Antarctic ice cover was measured between 2003 and 2009, are supposed to survey the entire region continuously. Further geophysical field measurements should be conducted inland and at the ice boundary, using aircraft. New biogeochemical and biological sensors are also needed. In addition, international cooperation and communications between individual researcher groups, as well as between automatic stations in Antarctica and laboratories at home, must increase.

In the future, the objective of research in Antarctica must be to maximize the scientific yield and to minimize the human footprint.

Further Reading

Antarctic Treaty: https://www.ats.aq/e/antarctictreaty.html (accessed December 4, 2020).
Antarctic Treaty (member states): https://www.ats.aq/devAS/Parties?lang=e (accessed December 4, 2020).
Horizon Scan: https://www.scar.org/science/horizon-scan/overview/ (accessed December 3, 2020); www.uc.pt/tomenota/2014/082014/documentos/20140806_1 (accessed December 3, 2020).
IceCube: http://icecube.wisc.edu/ (accessed December 3, 2020).

Appendix

Bibliography and Sources

Archives

Archiv für deutsche Polarforschung, Alfred-Wegener-Institut, Bremerhaven.
Filchnerarchiv, Bayerische Akademie der Wissenschaften, Munich.
GStA Merseburg: Geheimes Staatsarchiv Merseburg, heute: GStA PK Berlin (Geheimes Staatsarchiv Preußischer Kulturbesitz Abteilung Merseburg, Berlin).
Leibniz-Institut für Länderkunde, Leipzig.
Thüringisches Staatsarchiv Altenburg.
Zentrales Staatsarchiv Potsdam, heute: BArch Potsdam (Bundesarchiv Potsdam).

Unpublished Sources

Aufruf–Eine neue deutsche Südpolarexpedition, 1911, Filchnerarchiv, Bayerische Akademie der Wissenschaften, Munich.
Auswärtiges Amt: June 11, 1938, Anlage Protokoll der Sitzung vom 11. Juni 1938, No. R 13428. Leibniz-Institut für Länderkunde, Leipzig, Ritscher estate, File AA1, Abt. AA.
Auswärtiges Amt: June 3, 1938, Vermerk zu 1.Abtlk.Sk 1.1981/38 geh., Leibniz-Institut für Länderkunde, Leipzig, Ritscher estate, File AA1, Abt. AA.
Barkow, Erich: Tagebuch des Meteorologen Erich Barkow an Bord der Deutschland (Deutsche Antarktische Expedition 1911/12 unter der Leitung von Wilhelm Filchner), Private possession Joester, Bremen.
Beauftragter für den Vierjahresplan: September 3, 1938, Göring an Deutsche Lufthansa, ST.M.Dev. 1075 g.Rs., Leibniz-Institut für Länderkunde, Leipzig, Ritscher estate, File Bh1, Abt. VJP.
Beauftragter für den Vierjahresplan: October 5, 1938, ST.M.Dev. 1241 g.Rs., Leibniz-Institut für Länderkunde, Leipzig, Ritscher estate, File Bh1, Abt. VJP.

Berliner Tageblatt, issue of March 21, 1911, Bundesarchiv Potsdam, 09. 01 AA, No. 37564, Bl. 165.
Björvik, Paul: Schilderung einer Reise in die Weddell-See mit der Filchnerschen Expedition 1911/13 von Paul Björvik. Translation in Archiv für deutsche Polarforschung, Alfred-Wegener-Institut, Bremerhaven.
Bludau, Josef: o. D. Ärztlicher Bericht, Leibniz-Institut für Länderkunde, Leipzig, Ritscher estate, File Bh1, Abt. Medizin.
Deutsche Antarktische Expedition: August 24, 1939, Leibniz-Institut für Länderkunde, Leipzig, Ritscher estate, File Bh1, Abt. OKM.
Deutsche Antarktische Expedition: June 8, 1939, Entwurf eines Arbeitsprogramms der DAE 1939/40, Leibniz-Institut für Länderkunde, Leipzig, Ritscher estate, File Bh1, Abt. VJP.
Deutsche Forschungsgemeinschaft: December 8, 1938, Deutsche Forschungsgemeinschaft, Vermerk, R73 No. 242, Bl. 7–8, Bundesarchiv, Koblenz.
Deutsche Lufthansa: November 1, 1940, Deutsche Lufthansa, Leibniz-Institut für Länderkunde, Leipzig, Ritscher estate, File Bh2, Abt. VJP.
Deutsche Lufthansa: November 12, 1938, Brief an Dornier, Leibniz-Institut für Länderkunde, Leipzig, Ritscher estate, File AuE1, Abt. D.
Drygalski, Erich von: April 5, 1898, Brief an Hans Meyer, Private possession Kerler, Söcking.
Drygalski, Erich von: October 9, 1900, Bestellung, Leibniz-Institut für Länderkunde, Leipzig, Box 64, Inventory Number 4767, Serial Number 1.
Drygalski, Erich von: May 7, 1901, Brief an das Reichsministerium des Innern, BArch Potsdam, 15. 01. RMdI, No. 16120, Bl. 103–104.
Drygalski, Erich von: July 15, 1901, Brief an das Reichsministerium des Innern, GStA Merseburg, Rep. 76 Vc, Sekt.1, Tit. 11, Part VA, No. 7, Vol. I, Bl. 396–402.
Drygalski, Erich von: 1901–1903a, Stichwortartiger Auszug aus Drygalskis Südpolartagebüchern, Private possession Gazert, Partenkirchen.
Drygalski, Erich von: 1901–1903b, Auszug aus Drygalskis Südpolartagebüchern, Private possession Gazert, Partenkirchen.
Drygalski, Erich von: January 1, 1902, Brief an das Reichsministerium des Innern, GStA Merseburg, Rep. 76 Vc, Sect. 1, Title. 11, Part VA, No. 7, Vol. I, Bl. 410–417.
Drygalski, Erich von: January 25, 1902, Brief an das Reichsministerium des Innern, GStA Merseburg, Rep. 76 Vc, Sect. 1, Title 11, Part VA, No. 7, Vol. I, Bl. 418–422.
Drygalski, Erich von: May 30, 1903, Brief das Reichsministerium des Innern, BArch Potsdam, 15. 01. RMdI, No. 16148.
Drygalski, Erich von: June 1, 1903, Brief das Reichsministerium des Innern, BArch Potsdam, 15. 01. RMdI, No. 16127, Bl. 87.
Drygalski, Erich von: (1948), Unveröffentlichte Autobiographie, Private possession Mörder, Feldkirchen-Westerham.
Erprobungsstelle der Luftwaffe: October 29, 1938, Erprobungsstelle der Luftwaffe, Programm für Auftrag E728/11, Leibniz-Institut für Länderkunde, Leipzig, Ritscher estate, File AuE1, Abt. E.

Filchner, Wilhelm: (1900), Brief an Geheimes Civilcabinet, Geheimes Staatsarchiv Merseburg, 2.2.1. Geheimes Zivilkabinett, No. 21374, Bl. 5–7.

Filchner, Wilhelm: (end of 1900), Bewerbung, Leibniz-Institut für Länderkunde, Leipzig, Box 98, Inventory Number 6240, Serial Number 2.

Filchner, Wilhelm: October 4, 1911, Verabredungen für die Hilfsexpedition, Bundesarchiv Potsdam, 09. 01. AA, No. 37565, Bl. 100–104.

Hartmann, Gertraude: 2007, Auszug aus der Ahnenforschung von Alfred Ritscher, überarbeitet und ergänzt von seiner Tochter Gertraude Hartmann, geb. Ritscher, unpublished, Private possession Hartmann, Braunfels.

Hartmann, Gertraude: 2014, 1926 Tagebuch Südamerika-Vorbereitungen »Luther-Flug«, aufbereitet und recherchiert für ihre Familie von Gertraude Hartmann, geb. Ritscher, unpublished, Private possession Hartmann, Braunfels.

Kommission für die Beratung einer Südpolarexpedition: September 5, 1898, Sitzungsprotokoll, Leibniz-Institut für Länderkunde, Leipzig, Box 61, Inventory Number4753, Serial Number 1.

Kommission für die Beratung einer Südpolarexpedition: November 19, 1898, Sitzungsprotokoll, Leibniz-Institut für Länderkunde, Leipzig, Box 61, Inventory Number 4753, Serial Number 1.

Kommission für die deutsche Südpolarexpedition: June 27, 1899, Sitzungsprotokoll, Leibniz-Institut für Länderkunde, Leipzig, Box 61, Inventory Number 4753, Serial Number 1.

Kommission für die deutsche Südpolarexpedition: May 24, 1901, Verhandlungsprotokoll an Bord der Gauss, BArch Potsdam, 15.01, RMdI, No. 16120, Bl. 198–201.

Krümmel, Otto: April 15, 1901a, Brief an Drygalski, Leibniz-Institut für Länderkunde, Leipzig, Box 61, Inventory Number 4755, Serial Number 3.

Krümmel, Otto: April 15, 1901b, Die Trift des »Gauss« in den antarktischen Gewässern. Eine Prognose von O. Krümmel, Leibniz-Institut für Länderkunde, Leipzig, Box 61, Inventory Number 4755, Serial Number 3.

Larsen, Carl Anton: (1913), Vertraulicher Bericht, Auswärtiges Amt Berlin, AA VI B, Vol. 22, IIId 9021.

Lehmann: December 10, 1938, Aktennotiz, Leibniz-Institut für Länderkunde, Leipzig, Ritscher estate, File AA.

Norddeutscher Lloyd: March 26, 1903, Brief an das Reichsministerium des Innern, BArch Potsdam, 15. 01 RMdI, No. 16155.

Oberkommando der Kriegsmarine: August 4, 1938, Zur Besprechung vom 15. Juli 1938, B.No. 420/38 geh. BH V, Leibniz-Institut für Länderkunde, Leipzig, Ritscher estate, File Bh1, Abt. OKM.

Oberkommando der Kriegsmarine: August 11, 1938, B.No. 420/58 geh. BH VII Ang., Leibniz-Institut für Länderkunde, Leipzig, Ritscher estate, File Bh1, Abt. OKM.

Oberkommando der Kriegsmarine: November 3, 1938, B.No. 2078 g. Kds. W V, Anlage, Leibniz-Institut für Länderkunde, Leipzig, Ritscher estate, File Bh1, Abt. OKM.

Oberkommando der Kriegsmarine: November 11, 1938, B.No. 1087/37 g. Kds. BH W V, III. Ang., Leibniz-Institut für Länderkunde, Leipzig, Ritscher estate, File Bh1, Abt. OKM.

Oberkommando der Kriegsmarine: November 21, 1938, B.No. 2215/38 g. Kds. BH W V, Leibniz-Institut für Länderkunde, Leipzig, Ritscher estate, File Bh1, Abt. OKM.

Penck, Albrecht: June 9, 1913, Brief an Herzog Ernst von Altenburg. Thüringisches Staatsarchiv Altenburg, Haus- und Privatarchiv der Herzöge von Sachsen-Altenburg, No. 226 h.

Posadowsky-Wehner, Graf Arthur von: November 2, 1898, Brief an Thielmann, GStA Merseburg, Rep. 92, Estate of Althoff, Abt. B, No. 24, Vol. 3, Bl. 113–114, 125–126.

Posadowsky-Wehner, Graf Arthur von: July 27, 1901a, Brief an Drygalski, Leibniz-Institut für Länderkunde, Leipzig, Box 466, Akte d, Serial Number 98.

Posadowsky-Wehner, Graf Arthur von: July 27, 1901b, Geheime Ordre, Leibniz-Institut für Länderkunde, Leipzig, Box 466, File d, Serial number 109.

Posadowsky-Wehner, Graf Arthur von: June 2, 1903, Posadowsky an Drygalski, BArch Potsdam, 15.01 R MdI, No. 16127, Bl. 89.

Posadowsky-Wehner, Graf Arthur von: January 11, 1904, Brief an den Kaiser, GStA Merseburg, 2.2.1. Geheimes Zivilkabinett, No. 21373, Bl. 124.

Posadowsky-Wehner, Graf Arthur von: April 14, 1904, Aktennotiz, GStA Merseburg, 2.2.1. Geheimes Zivilkabinett, No. 21373, Bl. 138–141.

Regula, Herbert: February 2, 1939, Wetterberatung für einen Flug von der »Schwabenland« in südöstlicher Richtung, Leibniz-Institut für Länderkunde, Leipzig, Ritscher estate, File Bb1 Meteorologie.

Reichsministerium des Innern: March 10, 1899, Aktennotiz, BArch Potsdam, 15.01 RMdI, No. 16133, Bl. 4–6.

Reichsministerium des Innern: April 4, 1900, Vertrag mit den Howaldtswerken, BArch Potsdam, 15.01 RMdI, No. 16119, Bl. 27–30.

Reichsministerium des Innern: July 18, 1901a, Dienstanweisung für die »Deutsche Südpolar-Expedition«, Leibniz-Institut für Länderkunde, Leipzig, Box 70, Inventory Number 4870, Serial Number 6.

Reichsministerium des Innern: July 18, 1901b, Dienstanweisung für die »Deutsche Südpolar-Expedition« (Schiffsmannschaft), BArch Potsdam, 15.01 RMdI, No. 16121, Bl. 133–141, 231–233.

Reichsministerium des Innern: July 27, 1901, Brief an Drygalski, Leibniz-Institut für Länderkunde, Leipzig, Box 467, Akte k, Serial Number 185.

Reichsministerium des Innern: August 24, 1901, Brief an das Auswärtige Amt, BArch Potsdam, 15.01 RMdI, No. 16122, Bl. 86–87.

Reichsministerium des Innern: June 11, 1902, Brief an das Auswärtige Amt, BArch Potsdam, 15.01 RMdI , No. 16154, Bl. 101–106.

Reichsministerium des Innern: October 14, 1902, Entwurf der vierten Denkschrift, BArch Potsdam, 15.01 RMdI, No. 16137, 3 Bl.

Reichsministerium des Innern: May 6, 1903, Brief an Tafel, BArch Potsdam, 15.01 RMdI, No. 16155.

Appendix

Reichsministerium des Innern: June 23, 1903, Verhandlung im Reichsministerium des Innern, BArch Potsdam, 15.01. RMdI, No. 16127, Bl. 192–197.

Reichsministerium des Innern: July 11, 1903, Telegramm an Deutsche Südpolar-Expedition, Leibniz-Institut für Länderkunde, Leipzig, Box 74, Inventory Number 4825, Serial Number 3.

Reichsministerium für Ernährung und Landwirtschaft: June 29, 1938, II B57201 II, Leibniz-Institut für Länderkunde, Leipzig, Ritscher estate, File AA1, Abt. AA.

Reichsschatzamt: August 13, 1901, Brief an das Reichsministerium des Innern, BArch Potsdam, 15.01 RMdI, No. 16122, Bl. 85.

Reichsschatzamt: September 27, 1901, Brief an die Reichshauptkasse, BArch Potsdam, 15.01 RMdI, No. 16136.

Ritscher, Alfred: August 31, 1938, Brief an Beauftragten für den Vierjahresplan, Minister zur besonderen Verwendung, Leibniz-Institut für Länderkunde, Leipzig, Ritscher estate, File Bh1, Abt. VJP.

Ritscher, Alfred: December 1, 1938, Leibniz-Institut für Länderkunde, Leipzig, Ritscher estate, File AuE1, Abt. B.

Ritscher, Alfred: (January 1939), Organisationsplan für die beabsichtigten Antarktisflüge. Leibniz-Institut für Länderkunde, Leipzig, Ritscher estate, File Bb1, Abt. Flüge.

Ritscher, Alfred: May 2, 1939, Brief an das Oberkommando der Kriegsmarine, Leibniz-Institut für Länderkunde, Leipzig, Ritscher estate, File Bb1, Abt. Trinidad.

Ritscher, Alfred: August 19, 1940, Konzept, Leibniz-Institut für Länderkunde, Leipzig, Ritscher estate, File VAE1, Abt. no title.

Ritscher, Alfred: August 26, 1940, Arbeitsprogramm für die DAE 1940-41, Leibniz-Institut für Länderkunde, Leipzig, Ritscher estate, File VAE1, Abt. no title.

Ritscher, Alfred: September 14, 1940, Arbeitsprogramm für die DAE 1940-41, Leibniz-Institut für Länderkunde, Leipzig, Ritscher estate, File VAE1, Abt. no title.

Ritscher, Alfred: September 26, 1940, Notiz, Leibniz-Institut für Länderkunde, Leipzig, Ritscher estate, File AK 2, Abt. G.

Ritscher, Alfred: October 30, 1940, Notiz, Leibniz-Institut für Länderkunde, Leipzig, Ritscher estate, File AK 2, Abt. K.

Ritscher, Alfred: October 2, 1941, Brief an Klebelsberg, Leibniz-Institut für Länderkunde, Leipzig, Ritscher estate, File AK 2, Abt. K.

Ritscher, Alfred: December 27, 1941, Notiz, Leibniz-Institut für Länderkunde, Leipzig, Ritscher estate, File AK 2, Abt. G.

Sauter, Siegfried: May 25, 1992, Brief an Cornelia Lüdecke, Private possession Lüdecke, Munich.

Schirmacher, Richardheinrich: January 20, 1939, Flugprotokoll, Leibniz-Institut für Länderkunde, Leipzig, Ritscher estate, File Bb1, Abt. Flüge.

Tafel, Albert: April 15, 1903, Brief an das Reichsministerium des Innern, BArch Potsdam, 15.01 RMdI, No. 16156.

Thielmann, Max von: March 15, 1899, Brief an das Reichsministerium des Innern, BArch Potsdam, 15.01 RMdI, No. 16133, Bl. 8.

Thielmann, Max von: September 13, 1901, Brief an Posadowsky, BArch Potsdam, 15.01 RMdI, No. 16125, Bl. 2–3.

Tirpitz, Alfred von: October 4, 1898, Tirpitz, Gutachten zum Immediatgesuch, GStA Merseburg, Rep. 92, Althoff, estate Abt. B, No. 24, Vol. 2, Bl. 19–22.

Tirpitz, Alfred von: June 17, 1899, Brief an das Reichsministerium des Innern, BArch Potsdam, 15.01 RMdI, No. 16117, Bl. 117.

Trott zu Solz, von: February 7, 1913, von Trott zu Solz an den Kaiser. Geheimes Staatsarchiv Merseburg, Geheimes Zivilkabinett, 2.2.1, No. 21374, Bl. 104–112.

Vierjahresplan: January 17, 1939, Notiz, Leibniz-Institut für Länderkunde, Leipzig, Ritscher estate, File Bh1, Abt. VJP.

Wagner, Hermann/Drygalski, Erich von: October 29, 1910, Gedächtnisprotokoll der Sitzung am 29. Oktober, 1910, Geheimes Staatsarchiv Merseburg, Rep. 92 Schmidt-Ott, Abt. B, No. 29, Vol. 2, Bl. 167–172.

Wissenschaftlicher Beirat: November 24, 1899, Verhandlungen im Reichsministerium des Innern, Leibniz-Institut für Länderkunde, Leipzig, Box 61, No. 4754, Serial Number 2.

Wohlthat, Helmuth: October 10, 1938, W XVII/91 g.Rs., Leibniz-Institut für Länderkunde, Leipzig, Ritscher estate, File Bh1, Abt. VJP.

Wohlthat, Helmuth: October 19, 1938, W XVII/106, Leibniz-Institut für Länderkunde, Leipzig, Ritscher estate, File Bh1, Abt. VJP.

Wohlthat, Helmuth: November 10, 1938, W XVII/163, Leibniz-Institut für Länderkunde, Leipzig, Ritscher estate, File Bh1, Abt. VJP.

Wohlthat, Helmuth: November 21, 1938, Brief an Ritscher und andere beteiligte Stellen, W XVII/175, Leibniz-Institut für Länderkunde, Leipzig, Ritscher estate, File Bh1, Abt. VJP.

Wohlthat, Helmuth: December 8, 1938, W XVII/200, R73 No. 242, Bl. 2, Bundesarchiv, Koblenz.

Wohlthat, Helmut: December 10, 1938, W XVII/203, Bundesarchiv Koblenz, R 73/242, BL. 6.

Wohlthat, Helmuth: January 17, 1939, Telegramm an Ritscher auf der Schwabenland, Leibniz-Institut für Länderkunde, Leipzig, Ritscher estate, File AA1, Abt. Norwegen.

Wohlthat, Helmuth: June 6, 1939, W XVII/596, Leibniz-Institut für Länderkunde, Leipzig, Ritscher estate, File Bh1, Abt. VJP.

Wohlthat, Helmuth: August 11, 1939, W XVII/742, Leibniz-Institut für Länderkunde, Leipzig, Ritscher estate, File Bh1, Abt. VJP.

Wohlthat, Helmuth: September 5, 1939, Protokoll der Besprechung vom 5. September, 1939, W XVII/797, Leibniz-Institut für Länderkunde, Leipzig, Ritscher estate, File Bh1, Abt. VJP. Abs.

Literature

Otto: Professor Adolf Hoel 80 Jahre alt, in: Polarforschung 27 (1957), pp. 51–53.

Ahlbrecht, Bernhard: Internationale Walfangabkommen, deutsches Walfanggesetz und Reichsförderung, in: Nicolaus, Peters (ed.): Der neue deutsche Walfang, Hamburg 1938, pp. 24–45.

Akademischer Alpenverein Berlin: Jahresbericht 1903/04–1910/11, Berlin 1904–1911.

Akademischer Alpenverein München (ed.): Jahresbericht des Akademischen Alpenvereins München, 1901/02–1909/10, München 1903–1911.

Akademischer Alpenverein München (ed.): Josef Enzensperger. Ein Bergsteigerleben, München, 1924.

Angenheister, Gustav G.: Geschichte des Samoa-Observatoriums von 1902 bis 1921, in: Birett, Herbert/Helbig, K./Kertz, Walter/Schmucker, U. (eds.), Zur Geschichte der Geophysik. Festschrift zur 50jährigen Wiederkehr der Gründung der Deutschen Geophysikalischen Gesellschaft, Berlin 1974, pp. 43–66.

Anonymous: Polarländer, in: Geographische Zeitschrift 8 (1902), p. 53.

Anonymous: Polarländer, in: Geographische Zeitschrift 10 (1904), p. 231.

Anonymous: Polarländer, in: Geographische Zeitschrift 15 (1909), p. 652.

Anonymous: Filchners geplante Südpolarexpedition, in: Globus 97 (1910) 15, pp. 229–231.

Anonymous: Notiz ohne Titel, in: Globus 98 (1910) 8, p. 131.

Anonymous: Die deutsche Antarktische Expedition, in: Zeitschrift der Gesellschaft für Erdkunde zu Berlin 46 (1911), pp. 268–272.

Anonymous: Die deutsche Südpolarexpedition, in: Illustrierte Zeitung No. 3534 from March 23, 1911, pp. 535–538.

Anonymous: Süd-Polargegenden, in: Geographische Zeitschrift 19 (1913), pp. 290, 409.

Anonymous: Die Insel der Moschus–Ochsen, in: Berliner Nachtausgabe, issue of May 23, 1938.

Anonymous: Stuttgarter Illustrierte, issue 37, September 10, 1939.

Anonymous: Antarktisforscher brachen Rekord, in: Hamburger Abendblatt, issue of March 10, 1980.

Anonymous: Das Deutsche Polar-Forschungs- und Versorgungsschiff, in: Hansa 118 (1981) 17, pp. 1217–1220.

Anonymous: Deutscher Physiker in der Antarktis tödlich verunglückt, in: Süddeutsche Zeitung, issue of December 23, 1983.

Anonymous: Loch in der Ozonschicht wird größer, in: Süddeutsche Zeitung, issue of June 30, 1986.

Anonymous: Wettlauf zur Antarktis: Präsent sein ist alles, in: Der Spiegel (1988) 27, pp. 132–134.

Antarktische Informations und Schutzgemeinschaft: Antarktis–das letzte Paradies? Koblenz 1986.

Backe, Herbert: Um die Nahrungsfreiheit Europas. Weltwirtschaft oder Großraum. Leipzig 1942.

Barkley, Erich: Die biologischen Arbeiten, in: Vorbericht über die Deutsche Antarktische Expedition 1938/39, Annalen der Hydrographie und Maritimen Meteorologie, VIII. Supplement, pp. 19–21.

Barkow, Erich: Vorläufiger Bericht über die meteorologischen Beobachtungen der Deutschen Antarktischen Expedition 1911/12, in: Veröffentlichungen des Preußischen Meteorologischen Instituts, No. 265, Abhandlungen Vol. 4 (1913) 11, pp. 7–8.

Barkow, Erich: Die Ergebnisse der meteorologischen Beobachtungen der DAE 1911/12 (posthumous), in: Veröffentlichungen des Preußischen Meteorologischen Instituts, No. 325. Abhandlungen Vol. 7 (1924) 6.

Barr, Susan: Norway—a consistent polar nation? Oslo 2003.

Baschin, Otto: Die Südpolar-Expedition, in: Zeitschrift der Gesellschaft für Erdkunde zu Berlin 36 (1901) 4, pp. 165–218.

Baschin, Otto: Notiz ohne Titel, in: Zeitschrift der Gesellschaft für Erdkunde zu Berlin 45 (1911), p. 497.

Baumert, Susanne: Zu Wasser, im Eis und durch den Äther, in: Sobiesiak, Monika/ Korhammer, Susanne (eds.): Neun Forscherinnen im ewigen Eis. Die erste Antarktisüberwinterung eines Frauenteams, Basel, Boston, Berlin 1994, pp. 111–124.

Behrmann, Walter: Polargebiete, in: Zeitschrift der Gesellschaft für Erdkunde zu Berlin 46 (1911), pp. 128–130.

Beiträge zur Flottennouvelle 1900, Berlin 1900.

Bergdoll, Udo: Noch ist alles eingefroren. Warum Minister Matthöfer darauf drängt, daß Bonn möglichst bald dem »antarktischen Club« beitritt, in: Süddeutsche Zeitung, issue of February 13, 1978.

Berkman, Paul Arthur/Lang, Michael A./Walton, David W. H./Young, Oran R. (eds.): Science Diplomacy: Antarctica, Science, and the Governance of International Spaces, Washington, DC, 2011.

Bidlingmaier, Friedrich: Die erdmagnetisch-meteorologischen Arbeiten und Ausrüstungsgegenstände der deutschen Südpolar-Expedition und die Vorschläge für die internationale Kooperation während der Zeit der Südpolar-Forschung 1901–1903, in: Petermanns Geographische Mitteilungen 47 (1901), pp. 152–153.

Bidlingmaier, Friedrich: Zum ewigen Eise des Südpolar-Kontinents, in: Deutsches Knabenbuch 18 (1904), pp. 50–71.

Böhm, Ekkehard: Überseehandel und Flottenbau. Hanseatische Kaufmannschaft und deutsche Seerüstung, Düsseldorf 1972.

Brennecke, Wilhelm: Ozeanographische Arbeiten der Deutschen Antarktischen Expedition. I.–III. Bericht, in: Annalen der Hydrographie und Maritimen Meteorologie 39 (1911), pp. 350–353, 464–471, 642–647.

Brennecke, Wilhelm: Die ozeanographischen Arbeiten der Deutschen Antarktischen Expedition 1911–1912, in: Archiv der Deutschen Seewarte, 39 (1921), pp. 1–192.

Broszat, Martin/Frei, Norbert (eds.): Ploetz. Das Dritte Reich. Ursprünge, Ereignisse, Wirrungen, Freiburg, Würzburg 1983.

Bruce, William, Speirs: A new Scottish expedition to the South Polar regions, in: The Scottish Geographical Magazine (1908), pp. 200–202.

Bruns, Herbert: Deutsche Südpol-Expedition 1956/58. Ein akutes Problem der Gegenwart, München 1956.

Buddenbrock, Friedrich Frhr. von: »Atlantico« »Pazifico«. Lehrjahre des überseeischen Luftverkehrs, Düsseldorf 1965.

Bundesministerium für Umwelt, Naturschutz und Reaktorsicherheit (ed.): Deutsches Engagement für den weißen Kontinent. Antarktisvertrag 30 Jahre Konsultativstatus, Berlin 2011, p. 33. https://www.fona.de/mediathek/pdf/broschuere_antarktis_ de_bf2014.pdf (accessed November 26, 2019).

Creutzburg, Nikolaus: Erich von Drygalski zum 65. Geburtstag, in: Geographischer Anzeiger 26 (1925), pp. 1–8.

Degener, Herrmann August Ludwig (ed.): Wer ist's? Leipzig (1935), p. 345.

Denkschrift über die Deutsche Antarktische Expedition, Berlin 1911.

Deutscher Kolonialdienst, issue of May 15, 1938.

Drygalski, Erich von: Die Geoid-Deformation der Kontinente zur Eiszeit und ihr Zusammenhang mit den Wärmeschwankungen in der Erdrinde, Dissertation, in: Zeitschrift der Gesellschaft für Erdkunde zu Berlin 22 (1887), pp. 168–280.

Drygalski, Erich von: Über Bewegungen der Kontinente zur Eiszeit und ihren Zusammenhang mit den Wärmeschwankungen in der Erdrinde, in: Verhandlungen des 8. Deutschen Geographen-Tages zu Berlin, Berlin (1889), pp. 162–180.

Drygalski, Erich von: Die Südpolarforschung und die Probleme des Eises, in: Verhandlungen des 11. Deutschen Geographen-Tages in Bremen im Jahr 1895, Berlin 1896, pp. 18–30.

Drygalski, Erich von: Grönland-Expedition der Gesellschaft für Erdkunde zu Berlin 1891 bis 1893, 2 volumes, Berlin 1897.

Drygalski, Erich von: Die Aufgaben der Forschung an Nordpol und Südpol, in: Geographische Zeitschrift 4 (1898), pp. 121–133.

Drygalski, Erich von: Plan einer Deutschen Expedition in das Südpolargebiet, in: 7. Jahresbericht der Geographischen Gesellschaft Munich (1898), pp. 38–40.

Drygalski, Erich von, Plan und Aufgaben der Deutschen Südpolar-Expedition, in: Verhandlungen der Gesellschaft für Erdkunde zu Berlin 26 (1899), pp. 631–642.

Drygalski, Erich von: Deutschlands geographische Lage zur See, in: Beiträge zur Flottennouvelle 1900, Berlin 1900, pp. 76–93.

Drygalski, Erich von: Zum Kontinent des eisigen Südens, Berlin 1904. Abbreviated reprint; Drygalski, Erich von: Zum Kontinent des eisigen Südens: Die erste deutsche Südpolarexpedition 1901–1903, ed. by Cornelia Lüdecke, Wiesbaden 2013.

Drygalski, Erich von: Allgemeiner Bericht über die Arbeiten der Deutschen Südpolar-Expedition und deren Verwertung, in: Kollm, Georg (ed.): Verhandlungen des 15. Deutschen Geographen-Tages zu Danzig am 13., 14. und 15. Juni 1905, Berlin 1905, pp. 3–13.

Drygalski, Erich von (ed.): Deutsche Südpolar-Expedition 1901–1903 im Auftrage des Reichsamtes des Innern, 20 volumes, 2 atlases, Berlin, 1905–1931.

Drygalski, Erich von: Geographische Nachrichten, in: Mitteilungen der Geographischen Gesellschaft in München (1913), pp. 54–57.

Drygalski, Erich von: Buchbesprechung von A. Ritscher (1942), in: Geographische Zeitschrift 49 (1943) 6, p. 284.

Drygalski, Erich von: Entdeckungen und Ansprüche in der Antarktis, in: Geographische Zeitschrift 50 (1944) 1/2, pp. 55–63.

Drygalski, Erich von: Aus dem nachgelassenen Lebensrückblick, in: Mitteilungen der Geographischen Gesellschaft in München 75 (1990), pp. 119–141.

Edwards, Kerry/Graham, Robyn (ed.): Gender on ice, in: Proceedings of a conference on women in Antarctica held in Hobart, Tasmania, under the auspices of the Australian Antarctic Foundation, 19–21 August 1993, Canberra 1994.

Enß, Dietrich: Die deutsche Antarktis-Forschungsstation, in: Hansa 118 (1981) 13, pp. 963–966, 1011–1015.

Enzensperger, Josef J.: Die deutsche Südpolarexpedition, in: Petermanns Geographische Mitteilungen 48 (1902), pp. 68–71.

Enzensperger, Josef J.: Reisebriefe und Kerguelen-Tagebuch, in: Akademischer Alpenverein München (ed.): Josef Enzensperger. Ein Bergsteigerleben (posthumous), München 1905, pp. 209–276.

Fels, Edwin: Erich Dagobert v. Drygalski, in: Neue Deutsche Biographie, Berlin (1959), pp. 143–144.

Filchner, Wilhelm: Ein Ritt über den Pamir, Berlin, 1903.

Filchner, Wilhelm: Plan einer deutschen antarktischen Expedition, in: Zeitschrift der Gesellschaft für Erdkunde zu Berlin 45 (1910) 3, pp. 153–158.

Filchner, Wilhelm: Die Deutsche Antarktische Expedition, in: Zeitschrift der Gesellschaft für Erdkunde zu Berlin 45 (1910) 7, pp. 423–430.

Filchner, Wilhelm: Bericht des Expeditionsleiters Dr. Wilhelm Filchner, in: Zeitschrift der Gesellschaft für Erdkunde zu Berlin 47 (1912), pp. 83–90.

Filchner, Wilhelm: Zum sechsten Erdteil. Die zweite deutsche Südpolar-Expedition, Berlin 1922.

Filchner, Wilhelm: Ein Forscherleben, Wiesbaden 1950.

Filchner, Wilhelm: Feststellungen, in: Gottlob Kirschmer: Dokumentation über die Antarktisexpedition 1911/12 von Wilhelm Filchner. Deutsche Geodätische Kommission, München, Vol. E 23 (1985), pp. 24–58.

Filchner, Wilhelm: Aus den Tagebüchern von Wilhelm Filchner, in: Gottlob Kirschmer: Dokumentation über die Antarktisexpedition 1911/12 von Wilhelm Filchner, Deutsche Geodätische Kommission, München, Vol. E 23 (1985), pp. 89–120.

Filchner, Wilhelm: Om mani padme hum. Meine China- und Tibetexpedition, Reprint, Wiesbaden 2013.

Filchner, Wilhelm/Seelheim, Heinrich: Quer durch Spitzbergen. Eine deutsche Übungsexpedition im Zentralgebiet östlich des Eisfjords, Berlin 1911.

Fleischmann, Klaus: Zu den Kältepolen der Erde. 50 Jahre deutsche Polarforschung, Bielefeld 2005.

Frank, Wolfgang: Der wiedererstandene deutsche Walfang. Dargestellt an der Entwicklungsgeschichte der ersten deutschen Walfang-Gesellschaft in Verbindung mit einem Reisebericht über die 2. »Jan-Wellem«-Expedition, Düsseldorf 1939.

Fricker, Karl: Der VII. Internationale Geographenkongreß zu Berlin. Polarforschung, in: Geographische Zeitschrift 6 (1900), pp. 38–47.

Friederichsen, Ludwig: Der sechste Internationale Geographen-Kongreß in London 26. Juli–3. August 1895, in: Mitteilungen der Geographischen Gesellschaft Hamburg 1895, pp. 1–28.

Fritzsche, Diedrich: Geowissenschaftliche Forschung der DDR in der Antarktis, in: Schriftenreihe für Geowissenschaften 18, Zur Geschichte der Geowissenschaften in der DDR—Part II, Ostklüne 2011, pp. 303–317.

Fritzsche, Diedrich/Gernandt, Hartwig/Foken, Thomas: In Memoriam Günter Skeib, in: Polarforschung 81 (2011) 2, pp. 127–128.

Führer durch das Museum für Meereskunde in Berlin, Berlin, 1907.

Gazert, Hans: Bakteriologische Aufgaben der deutschen Südpolar-Expedition, in: Petermanns Geographische Mitteilungen 47 (1901), pp. 153–155.

Gazert, Hans: Gesundheitsbericht, in: Veröffentlichung des Instituts für Meereskunde zu Berlin, Issue 5 (1903), pp. 46–54.

Gazert, Hans: Unser Leben im Polareis, in: Westermanns Illustrierte Deutsche Monatshefte, October 577 (1904), pp. 40–53.

Gazert, Hans: Proviant und Ernährung bei der Deutschen Südpolar-Expedition, in: Drygalski, Erich von (ed.): Deutsche Südpolar-Expedition 1901–1903 im Auftrage des Reichsamtes des Innern, Vol. 7, Issue 4, Berlin 1908, pp. 1–73.

Gazert, Hans: Ärztliche Erfahrungen und Studien auf der Deutschen Südpolar-Expedition, in: Drygalski, Erich von (ed.): Deutsche Südpolar-Expedition 1901–1903 im Auftrage des Reichsamtes des Innern, Vol. 7, Issue 4, Berlin 1914, pp. 297–352.

Gazert, Hans/Renner, Otto: Die Beriberifälle auf Kerguelen, in: Drygalski, Erich von (ed.): Deutsche Südpolar-Expedition 1901–1903 im Auftrage des Reichsamtes des Innern, Vol. 7, Issue 4, Berlin 1914, pp. 353–386.

Georgi, Johannes: Polarforscher Kapitän Alfred Ritscher, in: Polarforschung 32 (1962) 1/2, pp. 125–127.

Gernandt, Hartwig; Erlebnis Antarktis, Berlin 1984.

Godwin, Joscely: Arktos. The Polar Myth in Science, Symbolism, and Nazi Survival, London 1996.

Goeldel, Wilhelm von: Ueber Versuche, die Knochenregenerationsfähigkeit des Rippenperiostes nach Rippen-Resektion zu verhüten, Dissertation, Berlin 1911.

Graser, Nora: Kalte Füße inklusive. Mein Jahr in der Antarktis, Munich 2008.

Gravenhort, Gode: Die Antarktisexpedition 1981/82, in: Gode Gravenhort: Antarktis-Expedition 1981/1982 (Unternehmen »Eiswarte«), in: Berichte zur Polarforschung 6/'82 (1982), pp. 9–17.

Grotewahl, Max: Prof. Dr. Max Robitzsch, in: Polarforschung 22 (1952), p. 145.

Gruber, Otto von: Das Wohlthat-Massiv im Kartenbild, in: Ritscher, Alfred (ed.): Wissenschaftliche und fliegerische Ergebnisse der Deutschen Antarktischen Expedition 1938/39, Vol. 1, Leipzig, pp. 157–230.

Güth, Rolf: Von Revolution zu Revolution. Entwicklungen und Führungsprobleme der Deutschen Marine 1848–1918, Herford 1978.

Hallstein, Walter Peter: Bekanntmachung über die Bestätigung der bei der Entdeckung von »Neu-Schwabenland« im Atlantischen Sektor der Antarktis durch die Deutsche Antarktische Expedition 1938/39 erfolgten Benennungen geographischer Begriffe, in: Bundesanzeiger 4 (1952) 149, issue of August 5, 1952, pp. 1–2.

Headland, Robert Keith: A Chronology of Antarctic exploration. A synopsis of events and activities from the earliest times until the International Polar Years, 2007–09, London 2009.

Heim, Fritz: Wissenschaftliche Ergebnisse der II. deutschen Südpolarexpedition, in: Mitteilungen der Geographischen Gesellschaft in München (1914), pp. 509–510.

Hempel, Gotthilf: Zum Aufbau des Alfred-Wegener-Instituts für Polarforschung, in: Polarforschung 51 (1981) 2, pp. 239–249.

Hempel, Gotthilf: Forscher, die in die Kälte gehen, in: Die Zeit, issue of November 3, 1989.

Hempel, Gotthilf: Deutsche Beiträge zur europäischen Zusammenarbeit in der Polarforschung, in: Polarforschung, 60 (1990) 3, pp. 245–250.

Hempel, Gotthilf: Blühende Landschaften im Ewigen Eis, in: Polarforschung 79 (2009) 3, 181–191.

Herold, Werner: Alltag in der Eiswüste, in: GEO 1986 (3) 3, pp. 38–50.

Herrligkoffer, Karl Maria: Deutsche Südpol-Expedition, München 1956.

Herrmann, Ernst: Deutsche Forscher im Südpolarmeer, Berlin 1941.

Herrmann, Ernst: Die geographischen Arbeiten, in: Ritscher, Alfred (ed.): Wissenschaftliche und fliegerische Ergebnisse der Deutschen Antarktischen Expedition 1938/39, Vol. 1, Leipzig 1942, pp. 282–304.

Herrmann, Ernst: Mit dem Fieseler-Storch ins Nordpolarmeer, Berlin 1942.

Herrmann, Ernst: Tektonik und Vulkanismus in der Antarktis und den benachbarten Meeresteilen, in: Petermanns Geographische Mitteilungen 1948, pp. 1–11.

Hornik, Helmut/Lüdecke, Cornelia: Wilhelm Filchner and Antarctica, in: Lüdecke, Cornelia (ed.): Steps of Foundation of Institutionalized Antarctic Research. Proceedings of the 1st SCAR Workshop on the History of Antarctic Research, Munich 2–3 June 2005, in: Reports on Polar and Marine Research 560 (2007), pp. 52–63. http://epic.awi.de/27231/ (accessed November 26, 2019).

Hugo, Otto: Deutscher Walfang in der Antarktis, Oldenburg i. O. 1939.

Jahresbericht der Sektion Berlin des Deutschen und Oesterreichischen Alpenvereins für 1910, Berlin 1910.

Jung-Hüttl, Angelika: Kaum Chancen für Antarktis-Schätze, in: Süddeutsche Zeitung, issue of January 19, 1987.

Jung-Hüttl, Angelika: Forschen im Kühlschrank der Erde, in: Kosmos (1992) 11, pp. 57–63.

Kennicutt II, Mahlon/Chown, Steven L., e.g.: Six priorities for Antarctic Science, in: Nature 512 (August 7, 2014), pp. 23–25.

Kirschmer, Gottlob: Filchner, Wilhelm, in: Neue Deutsche Biographie, Vol. 5 (1961), pp. 145.

Kirschmer, Gottlob: Dokumentation über die Antarktisexpedition 1911/12 von Wilhelm Filchner, in: Deutsche Geodätische Kommission, München, Vol. E 23 (1985).

Kistenmacher, Gert: Im Eiskeller ins Geschäft kommen, in: Süddeutsche Zeitung, issue of January 14, 1983.

Klebelsberg, Raimund von: Formen- und gletscherkundliche Auswertung der Lichtbildaufnahmen, in: Ritscher, Alfred: Wissenschaftliche und fliegerische Ergebnisse der Deutschen Antarktischen Expedition 1938/39, Vol. 1, Leipzig 1942, pp. 126–156.

Knaurs Konversationslexikon, Berlin 1934.

Kneißl, Max: Wilhelm Filchner zum Gedächtnis, in: Zeitschrift für Vermessungswesen 82 (1957) 9, pp. 314–320.

Knitschky, Wilhelm Ernst: Die Seegesetzgebung des Deutschen Reiches. Nebst den Entscheidungen des Reichsoberhandelsgerichts, des Reichsgerichts und der Seeämter, in: Guttentag'sche Sammlung Deutscher Reichsgesetze 19, 2. extended and improved edition, Berlin 1894.

Kohlberg, Eberhard: Eberhard Kohlberg, der einen Koch für die Antarktis sucht, in: Süddeutsche Zeitung, issue of January 31/February 1, 2015.

Kohl-Larsen, Ludwig: An den Toren der Antarktis, Stuttgart 1930.

Kohnen, Heinz: Antarktis Expedition. Deutschlands neuer Vorstoß ins ewige Eis, Bergisch Gladbach 1981.

Kollm, Georg: Verhandlungen des 13. Deutschen Geographentages zu Breslau am 28., 29. und 30. Mai 1901, Berlin 1901.

Kollm, Georg: Verhandlungen des VII. Internationalen Geographen-Kongresses zu Berlin von 28.9.–4.10.1899, 2 volumes, Berlin 1901.

Kosack, Hans-Peter: Die Neubearbeitung der Übersichtskarte des Arbeitsgebietes der Expedition, in: Ritscher, Alfred (ed.): Deutsche Antarktische Expedition 1938/39, Wissenschaftliche Ergebnisse Vol. 2, Issue 1, Hamburg 1954, pp. 1–15.

Kraul, Otto: Käpt'n Kraul erzählt. 20 Jahre Walfänger unter argentinischer, russischer und deutscher Flagge in der Arktis und Antarktis, Berlin 1939.

Krause, Artur Bernhard: Organisation von Arbeit und Wirtschaft, Berlin 1935.

Krause, Reinhard A.: Hintergründe der deutschen Polarforschung von den Anfängen bis heute, in: Deutsches Schiffahrtsarchiv 16 (1993), pp. 7–70.

Krause, Reinhard/A. Ursula Rack (eds.): Schiffstagebuch der Steam-Bark Groenland geführt auf einer Fangreise in die Antarktis im Jahre 1873/1874 unter der Leitung von Capitain Ed. Dallmann, in: Berichte zur Polar- und Meeresforschung 530 (2006).

Krause, Reinhard A./Salewski, Christian, R.: Das Alfred-Wegener-Institut Helmholtz-Zentrum für Polar- und Meeresforschung. Chronik und wissenschaftshistorische Grundlagen. Bremerhaven (2013/14).

Kretschmer (Marine-Oberbaurat): Die Südpolarexpedition, Berlin 1900.

Kretzer, Hans-Jochen: Windrose und Südpol, Leben und Werk des großen Pfälzer Wissenschaftlers Georg von Neumayer, Bad Dürkheim 1984.

Kuczynski, Thomas: Das Wachstum der Industrieproduktion in den kapitalistischen Hauptländern (England, USA, Frankreich, Deutschland) und seine regionale Verteilung von 1830 bis 1913. Versuch einer statistischen Rekonstruktion, in: Jahrbuch für Wirtschaftsgeschichte, Sonderband Umwälzung der deutschen Wirtschaft im 19. Jahrhundert (1989), pp. 183–191.

Lange, Annemarie: Das wilhelminische Berlin. Zwischen Jahrhundertwende und Novemberrevolution, Berlin 1988.
Lange, Gerd: Bewährung in Antarktika. Antarktisforschung der DDR, Leipzig 1982.
Lange, Gerd: Sonne, Sturm und weiße Finsternis. Eine Chronik der ostdeutschen Antarktisforschung, Hamburg 1996.
Lange, Heinz: Die Arbeiten der Expeditionswetterwarte. Part II: Radiosondenaufstiege, in: Vorbericht über die Deutsche Antarktische Expedition 1938/39, in: Annalen der Hydrographie und Maritimen Meteorologie VIII. Supplement 1939, pp. 35–36.
Lohmann, Hans: Bericht über die biologischen Arbeiten auf der Fahrt nach Buenos-Aires, in: Zeitschrift der Gesellschaft für Erdkunde zu Berlin (1912), pp. 94–101.
Lohmann, Hans: Untersuchungen über das Pflanzen- und Tierleben der Hochsee, zugleich ein Bericht über die biologischen Arbeiten auf der Fahrt der »Deutschland« von Bremerhaven bis Buenos-Aires in der Zeit vom 7. Mai bis 7. September, in: Veröffentlichungen des Instituts für Meereskunde an der Universität Berlin, Neue Folge, A. Geographisch-naturwissenschaftliche Reihe 1 (1912), pp. 1–92.
Louis, Herbert: Die Geographische Gesellschaft München, Rückblick im hundertsten Jahre ihres Bestehens, in: Mitteilungen der Geographischen Gesellschaft in München 54 (1969), pp. 5–20.
Lübke, Anton: Das deutsche Rohstoffwunder. Wandlungen der deutschen Rohstoffwirtschaft, Stuttgart 1943.
Lüdecke, Cornelia: Die Routenfestlegung der ersten deutschen Südpolarexpedition durch Georg von Neumayer und ihre Auswirkung, in: Polarforschung 59 (1989), pp. 103–111.
Lüdecke, Cornelia: Vor 100 Jahren: Grönlandexpedition der Gesellschaft für Erdkunde zu Berlin (1891, 1892–1893) unter der Leitung Erich von Drygalskis, in: Polarforschung 60 (1990), pp. 219–229.
Lüdecke, Cornelia: Die erste deutsche Südpolar-Expedition und die Flottenpolitik unter Kaiser Wilhelm II, in: Historisch-meereskundliches Jahrbuch 1 (1992), pp. 55–75.
Lüdecke, Cornelia: Die deutsche Polarforschung seit der Jahrhundertwende und der Einfluß Erich von Drygalskis, Dissertation, in: Berichte zur Polarforschung 158 (1995). http://hdl.handle.net/10013/epic.10159 (accessed November 26, 2019).
Lüdecke, Cornelia: Ein Meeresstrom über dem Südpol? Vorstellungen von der Antarktis um die Jahrhundertwende, in: Historisch-meereskundliches Jahrbuch 3 (1995), pp. 35–50.
Lüdecke, Cornelia: Zum 100. Geburtstag von Max Grotewahl (1894–1958), Gründer des Archivs für Polarforschung, in: Polarforschung 65 (1995), pp. 93–105.
Lüdecke, Cornelia: Erich von Drygalski und der Aufbau des Instituts und Museums für Meereskunde, in: Historisch-meereskundliches Jahrbuch 4 (1997), pp. 19–36.
Lüdecke, Cornelia: Scientific collaboration in Antarctica (1901–1903): a challenge in times of political rivalry, in: Polar Record 39 (2003), pp. 25–48.

Lüdecke, Cornelia: In geheimer Mission zur Antarktis. Die dritte Deutsche Antarktisexpedition 1938/39 und der Plan einer territorialen Festsetzung zur Sicherung des Walfangs, in: Deutsches Schiffahrtsarchiv 26 (2003), pp. 75–100.

Lüdecke, Cornelia: Karl Maria Herrligkoffer's private "German South Pole Expedition" 1957/58: A failed initiative, in: Lüdecke, Cornelia (ed.): Steps of Foundation of Institutionalized Antarctic Research. Proceedings of the 1st SCAR Workshop on the History of Antarctic Research, Munich 2–3 June 2005, in: Reports on Polar and Marine Research 560 (2007), pp. 195–210.

Lüdecke, Cornelia: Diverging Currents—Depicting Southern Ocean Currents in the Early 20th Century, in: Keith R. Benson/Helen M. Rozwadowski (eds.): Extremes: Oceanography's Adventures at the Poles, Sagamore Beach 2007, pp. 71–105.

Lüdecke, Cornelia: International Cooperation in Antarctica (1901–1904), in: Barr, Susan/Lüdecke, Cornelia (eds.): The History of the International Polar Years (IPYs). Berlin, Heidelberg (2010), pp. 127–134.

Lüdecke, Cornelia: Gorgeous Landscapes and Wildlife: The Importance and danger of Antarctic Tourisms. (Paisaches y vida silvestra maravillosa. Importancia y peligros del turismo Antárctico), in: Estudios Hermifericos y polares (Hemispheric & Polar Studies Journal) 1 (2010) 4, pp. 213–231. online under: http://www.hemisfericosypolares.cl/articulos/014-Luedecke-Importance%20Danger%20Antarctic%20Tourism.pdf (acceded 5 August 2020)

Lüdecke, Cornelia: Roald Amundsen. Ein biografisches Porträt, Freiburg 2011.

Lüdecke, Cornelia: Investigation of the unknown: The flight programme of the German "Schwabenland" expedition 1938/39, in: The Polar Journal 2 (2012) 2, pp. 312–333.

Lüdecke, Cornelia (ed.): Verborgene Eiswelten. Erich von Drygalskis Bericht über seine Grönlandexpeditionen 1891, 1892–1893, München, 2015.

Lüdecke, Cornelia/Brogiato, Heinz-Peter/Hönsch, Ingrid: Universitas Antarctica. 100 Jahre deutsche Südpolarexpedition 1901–1903 unter der Leitung Erich von Drygalskis, Leipzig 2001.

Lüdecke, Cornelia/Summerhayes, Colin: The Third Reich in Antarctica: The German Antarctic Expedition 1938–39, Eccles and Bluntisham, 2012.

Luyken, Karl: Die erdmagnetischen Arbeiten auf der Kerguelen-Station, in: Kollm, Georg (ed.): Verhandlungen des 15. Deutschen Geographentages zu Danzig am 13., 14. und 15. Juni 1905, Berlin 1905, pp. 57–64.

Mandelsloh, Klaus von/Freyenhagen, Jörn: Antarktis. Entdeckungsfahrten in die Zukunft, München 1982.

Markham, Clements: The Antarctic expeditions, in: Kollm, Georg (ed.): Verhandlungen des VII. Internationalen Geographen-Kongresses zu Berlin von 28.9.–4.10.1899, Vol. 2 (1899), pp. 623–630.

Markham, Clements: Antarctic Obsession. A personal narrative of the origins of the British national Antarctic expedition 1901–1904 (posthumous), edited by Clive Holland, Alburgh 1986.

Martin, Stephen: A History of Antarctica, 2nd edition, New South Wales, Australia 2013.

Mayr, Rudolf: Deutsche Flugboote fliegen über die Antarktis, in: Der Lufthanseat 3 (1939) 6, pp. 1–6.

Meier, Siegfried: 450 Tage in Antarktika, Leipzig 1975.

Meinardus, Wilhelm: Erich von Drygalski †, in: Petermanns Geographische Mitteilungen 93 (1949), pp. 177–180.

Messerschmidt, Manfred: Reich und Nation im Bewusstsein der wilhelminischen Gesellschaft, in: Schottelius, Herbert/Deist, Wilhelm (eds.): Marine und Marinepolitik im kaiserlichen Deutschland 1871–1914, Düsseldorf 1981, 2nd. edition, pp. 11–33.

Meynen, Emil: Deutscher Geographentag 1881–1963. Gesamtinhaltsverzeichnis der Verhandlungen, Wiesbaden, 1965.

Miethe, Adolf/Hergesell, Hugo (eds.): Mit Zeppelin nach Spitzbergen, Berlin 1911.

Miller, Heinz/Oerter, Hans (ed.): Die Ekström-Traverse 1987, in: Berichte zur Polarforschung 57/90 (1990), pp. 53–55.

Mitteilungen für die Vereinigung zur Förderung des Archivs für Polarforschung 1931.

Muhle, Grazyna/Korhammer, Susanne: Vom Dieselgenerator bis zur Spülmaschine, in: Sobiesiak, Monika/Korhammer, Susanne (eds.): Neun Forscherinnen im ewigen Eis. Die erste Antarktisüberwinterung eines Frauenteams, Basel, Boston, Berlin 1994, pp. 51–71.

Müller, Johannes: Reiseweg und Lotungen, in: Annalen der Hydrographie und maritimen Meteorologie (1912) Tafel 7.

Nansen, Fridtjof: In Nacht und Eis. Die norwegische Polarexpedition 1893–1896. Reprint, Wiesbaden (2011).

Neumayer, Georg: Die Erforschung des Süd-Polar-Gebietes, in: Zeitschrift der Gesellschaft für Erdkunde zu Berlin (1872), pp. 120–170.

Neumayer, Georg: Thätigkeitsbericht der Deutschen Kommission für die Südpolar-Forschung, in: Kollm, Georg (ed.): Verhandlungen des 12. Deutschen Geographen-Tages in Jena im Jahr 1897, Berlin 1897, pp. 15–29.

Neumayer, Georg von: Auf zum Südpol! 45 Jahre Wirkens zur Förderung der Erforschung der Südpolarregion 1855–1900, Berlin, 1901.

Neumayer, Georg von: Zweiter Thätigkeitsbericht der Deutschen Kommission für die Südpolar-Forschung, in: Kollm, Georg (ed.): Verhandlungen des 13. Deutschen Geographen-Tages zu Breslau am 28., 29. und 30. Mai 1901, Berlin 1901 pp. 3–32.

Noack, Gerold: Antarktis. Abenteuer Wissenschaft—Ein Lausitzer im ewigen Eis, Cottbus 2014.

Oberhummer, Eugen: Die Deutsche Südpolarexpedition. Bericht über die vorbereitenden Schritte und die Versammlung in München am 13. Mai 1898, in: 17. Jahresbericht der Geographischen Gesellschaft in München 1898, pp. 1–48.

Oberhummer, Eugen: Die Deutsche Südpolarexpedition. Zweiter Bericht der geographischen Gesellschaft in München, in: 18. Jahresbericht der Geographischen Gesellschaft in München 1900, pp. 94–134.

Oberhummer, Eugen: Die Deutsche Südpolarexpedition. Dritter Bericht der geographischen Gesellschaft in München, in: 19. Jahresbericht der Geographischen Gesellschaft in München 1901, pp. 99–133.

Oerter, Hans/Drücker, Cord/Kipfstuhl, Sepp/Wilhelms, Frank: Kohnen Station—the Drilling Camp for the EPICA Deep Ice Core in Dronning Maud Land, in: Polarforschung 78 (2008), pp. 1–23.

Orheim, Olav: How Norway got Dronning Maud Land, in: Winther, Jan-Gunnar (ed.): Norway in the Antarctic—from Conquest to Modern Science, Oslo 2008, pp. 44–59.

Paech, HansJürgen: Die DDR-Antarktisforschung—eine Retrospektive, in: Polarforschung 69 (1990) 3, pp. 197–218.

Paulsen, Karl-Heinz: Die ozeanographischen Arbeiten, in: Vorbericht über die Deutsche Antarktische Expedition 1938/39, in: Annalen der Hydrographie und Maritimen Meteorologie, V III, Supplement 1939, pp. 27–33.

Pentzlin, Heinz: Hjalmar Schacht. Leben und Wirken einer umstrittenen Persönlichkeit, Berlin 1980.

Personen und Vorlesungsverzeichnis für das Sommersemester 1947, Universität München, München 1947.

Personen und Vorlesungsverzeichnis für das Wintersemester 1947/48, Universität München, München 1947.

Peters, Nicolaus: Über Hochseewalfang und Tierleben im Südlichen Eismeer, in: Der Fischmarkt (1937) 7/8, pp. 1–28.

Peters, Nicolaus: Kurze Geschichte des Walfangs von den ältesten Zeiten bis heute, in: Peters, Nicolaus (ed.): Der neue deutsche Walfang, Hamburg 1938, pp. 6–23.

Petzina, Dietmar: Die deutsche Wirtschaft in der Zwischenkriegszeit, Wiesbaden 1977.

Philipp, Hans (ed.): Ergebnisse der W. Filchnerschen Vorexpedition nach Spitzbergen 1910, in: Petermanns Geographische Mitteilungen, Ergänzungsheft 179 (1914).

Philippi, Emil: Die Schlittenreisen der Deutschen Südpolarexpedition, in: Deutsche Revue 30 (1905), pp. 103–112.

Philippi, Emil: Über das Problem der Schichtung und über Schichtbildung am Boden der heutigen Meere, in: Zeitschrift der Deutschen Geologischen Gesellschaft 60 (1908), pp. 346–377.

Pietsch, Jani: »Ich besaß einen Garten in Schöneich bei Berlin«: Das verwaltete Verschwinden jüdischer Nachbarn und ihre schwierige Rückkehr, Frankfurt am Main 2006, pp. 123–135.

Poeck, Klaus: Neurologie, Berlin (1978).

Pogge von Strandmann, Hartmut: Nationale Verbände zwischen Weltpolitik und Kontinentalpolitik, in: Schottelius, Herbert/Deist, Wilhelm (eds.): Marine und Marinepolitik im kaiserlichen Deutschland 1871–1914, Düsseldorf 1981, 2nd edition, pp. 296–317.

Potpeschnigg, Karl: Verlauf und Ausrüstung der Expedition, in: Hans Philipp (ed.): Ergebnisse der W. Filchnerschen Vorexpedition nach Spitzbergen 1910, in: Petermanns Geographische Mitteilungen, Ergänzungsheft 179 (1914), pp. 1–13.

Przybyllok, Erich: Deutsche Antarktische Expedition. Bericht über die Tätigkeit nach Verlassen von Südgeorgien, in: Zeitschrift der Gesellschaft für Erdkunde zu Berlin (1913), pp. 1–17.

Puskeppeleit, Monika: The untold story: The German allfemale overwintering, in: Edwards, Kerry/Graham, Robyn (eds.): Gender on ice, in: Proceedings of a conference on women in Antarctica held in Hobart, Tasmania, under the auspices of the Australian Antarctic Foundation, 19–21 August 1993, Canberra 1994, pp. 49–52.

Puskeppeleit, Monika: The allfemale expedition: Personal perspective, in: Edwards, Kerry/Graham, Robyn (eds.): Gender on ice, in: Proceedings of a conference on women in Antarctica held in Hobart, Tasmania, under the auspices of the Australian Antarctic Foundation, 19–21 August 1993, Canberra 1994, pp. 75–81.

Puskeppeleit, Monika: Die wahren Abenteuer sind im Kopf, in: Sobiesiak, Monika/ Korhammer, Susanne (eds.): Neun Forscherinnen im ewigen Eis. Die erste Antarktisüberwinterung eines Frauenteams, Basel, Boston, Berlin 1994, pp. 125–146.

Rack, Ursula: Sozialhistorische Studie zur Polarforschung: anhand von deutschen und österreich-ungarischen Polarexpeditionen zwischen 1868–1939, Dissertation, Wien 2009. http://othes.univie.ac.at/7081/1/20090730_8303884.pdf (accessed November 26, 2019).

Regula, Herbert: Die Arbeiten der Expeditionswetterwarte. Part I: Terminbeobachtungen, Höhenwindmessungen, Wetterdienst, Sonderuntersuchungen, in: Vorbericht über die Deutsche Antarktische Expedition 1938/39, in: Annalen der Hydrographie und Maritimen Meteorologie, VIII. Supplement 1939, pp. 33–35.

Regula, Herbert: Die Wetterverhältnisse während der Expedition und die Ergebnisse der meteorologischen Messungen, in: Ritscher, Alfred (ed.): Deutsche Antarktische Expedition 1938/39. Wissenschaftliche Ergebnisse Vol. 2, Issue 1, Hamburg 1954, pp. 16–40.

Reincke-Kunze, Christine: Aufbruch in die weiße Wildnis. Die Geschichte der deutschen Polarforschung, Hamburg 1992.

Renner, Erich: Ludwig Kohl-Larsen–Der Mann, der Lucy's Ahnen fand. Lebenserinnerungen und Materialien, Landau/Pfalz 1991.

Rieche, Herbert: Bericht über die »Deutsche Spitzbergen-Expeditionen 1937 und 1938«, in: Petermanns Geographische Mitteilungen 85 (1929), pp. 125–127.

Ritscher, Alfred: Wanderung in Spitzbergen im Winter 1912, in: Zeitschrift der Gesellschaft für Erdkunde zu Berlin (1916), pp. 16–34.

Ritscher, Alfred: Neuland in der Antarktis, in: Münchner Neueste Nachrichten, issue of May 5, 1939.

Ritscher, Alfred (ed.): Wissenschaftliche und fliegerische Ergebnisse der Deutschen Antarktischen Expedition 1938/39, Leipzig 1942.

Ritscher, Alfred: Oasen in Antarktika, in: Polarforschung 16 (1946) 1/2, pp. 70–71.

Ritscher, Alfred: Vor 10 Jahren, in: Polarforschung 18 (1948) 1/2, pp. 30–32.

Ritscher, Alfred (ed.): Deutsche Antarktische Expedition 1938/39. Wissenschaftliche Ergebnisse, Vol. 2, Hamburg 1954.

Rose, Lisle A.: Assault on Eternity. Richard E. Byrd and the Exploration of Antarctica 1946–47, Annapolis, 1980.

Rüdiger, Hermann: Deutschlands Anteil an der Lösung der polaren Probleme, Dissertation, in: Mitteilungen der Geographischen Gesellschaft München VII (1912) 4, pp. 455–564.

Rüdiger, Hermann: Die Sorge-Bai. Aus den Schicksalstagen der Schröder-Stranz-Expedition, Berlin 1913.

Salomon, Wilhelm: Die Spitzbergenfahrt des Internationalen Geologischen Kongresses, in: Geologische Rundschau 1 (1910) 6, pp. 302–309.

Shackleton, Ernest Henry: The Imperial Trans-Antarctic Expedition, London 1914.

Schirmacher, Richardheinrich/Mayr, Rudolf: Flüge über der unerforschten Antarktis, in: Ritscher, Alfred (ed.): Wissenschaftliche und fliegerische Ergebnisse der Deutschen Antarktischen Expedition 1938/39, Vol. 1, Leipzig 1942, pp. 231–265.

Schmidt, Josef: Antarktisexpedition erleidet Schiffbruch. Deutsches Forschungsschiff im Packeis leckgeschlagen und gesunken, in: Süddeutsche Zeitung, issue of 19/20 December 1981.

Schmidt-Ott, Friedrich: Erlebtes und Erstrebtes 1860–1950, Wiesbaden 1952.

Schnall, Uwe: Staat und Seekartographie im wilhelminischen Deutschland, in: Lindgren, Uta (ed.): Kartographie und Staat. Algorismus 3 (1990), pp. 55–65.

Scholl, Lars-Uwe: German Whaling in the 1930s, in: Fischer, Lewis R./Norvik, Helge W./Minchinton, Walter E. (eds.): Shipping and trade in the Northern Seas 1600–1939. Yearbook of the Association for the History of the Northern Seas 1988, pp. 103–121.

Scholz, Arnulf: 5 Jahre »Archiv für Polarforschung«, in: Polarforschung 2 (1932), p. 2.

Schön, Heinz: Mythos Neu-Schwabenland. Für Hitler am Südpol, Selent 2004.

Schumacher, Arnold: Die Lotungen der »Schwabenland«, in: Ritscher, Alfred (ed.): Deutsche Antarktische Expedition 1938/39, Wissenschaftliche und fliegerische Ergebnisse, Vol. 2, Hamburg 1958, pp. 41–62.

Simonov, I. M./Stackebrandt, W./Haendl, D./Kaup, E./Kämpf, H./Loopmann, A.: Report on Scientific Investigations at the Untersee and Obersee Lake, Central Dronning Maud Land (East Antarctica), in: Geodätische und geophysikalische Veröffentlichungen Reihe I (1985), Issue 12, pp. 8–25.

Skeib, Günter: Orkane über Antarktika. Forscherarbeit in Schnee und Eis, Leipzig 1963.

Sobiesiak, Monika/Korhammer, Susanne (eds.): Neun Forscherinnen im ewigen Eis. Die erste Antarktisüberwinterung eines Frauenteams, Basel, Boston, Berlin 1994.

Spengemann, Herbert: Auf Walfang in der Antarktis, Bühl-Baden 1939.

Stocks, Theodor: Lotarbeiten der »Schwabenland« December 1938–April 1939, in: Vorbericht über die Deutsche Antarktische Expedition 1938/39, in: Annalen der Hydrographie und Maritimen Meteorologie, VIII. Supplement 1939, pp. 36–40.

Stocks, Theodor: In Memoriam Alfred Ritscher 1879–1963, in: Deutsche Hydrographische Zeitschrift 16 (1963) 3, pp. 87–92.

Summerhayes, Colin P./Beeching, Peter: Hitler's Antarctic base: the myth and the reality, in: Polar Record 43 (2007) 224, pp. 1–21.

Szabo, Ladislas: Hitler esta vivo. El Tabano, Buenos Aires, 1947.
Szabo, Ladislav, Je sai que Hitler est vivant. SFELT, Paris, 1947.
Tammiksaar, Erki/Suchova, Natal'ja G.: August Petermann und seine Hypothesen über das Nordpolarmeer, in: Polarforschung 65 (1995), pp. 133–143.
Thorbecke, Franz: Die deutsche Südpolar-Expedition, in: Geographische Zeitschrift 11 (1905), pp. 503–510.
Tiedemann Karl-Heinz/Ruthe, Kurt: 25 Jahre Archiv für Polar-forschung, in: Polarforschung 21 (1951), pp. 81–83.
Tilly, Richard H.: Vom Zollverein zum Industriestaat. Die wirtschaftlich soziale Entwicklung Deutschlands 1834–1914, Munich 1990.
Tirpitz, Alfred von: Erinnerungen. New revised edition, Leipzig 1920.
Tripphahn, Bodo: Aus der Werkstatt der Expeditionen, in: Lange, Gerd: Sonne, Sturm und weiße Finsternis. Eine Chronik der ostdeutschen Antarktisforschung, Hamburg 1996, pp. 275–283.
Ule, Willi: Bericht über geographische Studien, in: Zeitschrift der Gesellschaft für Erdkunde zu Berlin (1912), pp. 101–107.
Ule, Willi: Quer durch Süd-Amerika, Lübeck, 1924.
Verhandlungen: Gemeinschaftliche Sitzung der Gesellschaft für Erdkunde zu Berlin und der Abteilung Berlin-Charlottenburg der Deutschen Kolonialgesellschaft am 16. Januar 1899, in: Verhandlungen der Gesellschaft für Erdkunde zu Berlin 26 (1899), pp. 58–87.
Vorbericht: Vorbericht über die Deutsche Antarktische Expedition 1938/39, in: Annalen der Hydrographie und Maritimen Meteorologie. VIII. Supplement 1939.
Wagner, Hermann: Besprechung »Zum Kontinent des eisige Südens«, in: Zeitschrift der Gesellschaft für Erdkunde zu Berlin (1905), pp. 331–347.
Walton, David W. H./Clarkson, Peter D.: Science in the Snow. Fifty years of international collaboration through the Scientific Committee on Antarctic Research, Cambridge 2011, 2nd extended edition 2018.
Wegener, Karl August: Die deutsche Kolonie in der Antarktis, in: Peters, Nicolaus (ed.): Der neue deutsche Walfang, Hamburg 1938, pp. 1–5.
Weiken, Karl: Prof. Dr. Bernhard Brockamps Verdienste um die deutsche Polarforschung und um die Deutsche Gesellschaft für Polarforschung, in: Polarforschung 38 (1968), pp. 187, 190–193.
Westphal, Wilfried: Geschichte der deutschen Kolonien, München, 1984.
Wichmann, Hugo: Südpolargebiete, in: Petermanns Geographische Mitteilungen 57 (1911), p. 84.
Wichmann, Hugo: Südpolargebiete, in: Petermanns Geographische Mitteilungen 56 (1910), pp. 29, 150, 210.
Wichmann, Hugo: Südpolargebiete, in: Petermanns Geographische Mitteilungen 59 (1913), p. 30.
Winterhoff, Edmund: Walfang in der Antarktis, Oldenburg 1974.
Witt, Peter-Christian: Die Finanzpolitik des Deutschen Reiches von 1903 bis 1913. Eine Studie zur Innenpolitik des Wilhelminischen Deutschland, Lübeck 1970.

Witt, Peter-Christian: Reichsfinanzen und Rüstungspolitik 1898–1914, in: Schottelius, Herbert/Deist, Wilhelm (eds.): Marine und Marinepolitik im kaiserlichen Deutschland 1871–1914, Düsseldorf 1981, 2nd edition, pp. 146–177.

Wohlthat, Helmuth: Die Deutsche Antarktische Expedition 1938/39, in: Der Vierjahresplan 3 (1939) 9, pp. 613–617.

Wohlthat, Helmuth: Walöl im Weltmarkt, in: Der Vierjahresplan 3 (1939) 11, pp. 726–731.

Wohlthat, Helmuth: Neue Entwicklungsmöglichkeiten des deutschen Verrechnungsverkehrs, in: Staatenwirtschaft, ständige Beilage zur Zeitschrift für Geopolitik 3 (1939) 4/5, pp. 701–706.

Internet Resources

7th GDR expedition group-14. SAE: January 1, 2000/PolarNEWS. http://antarktis.ch/2000/01/01/7ddrexpeditionsgruppe14sae/ (accessed November 27, 2019).

Alfred-Wegener-Institut (AWI): http://www.awi.de/en.html (accessed November 27, 2019).

Antarctic Treaty: https://www.ats.aq/e/antarctictreaty.html (accessed November 27, 2015).

Antarctic Treaty (member states): https://www.ats.aq/devAS/Parties?lang=e (accessed July 17, 2020).

AWI press release from February 20, 2009: https://www.awi.de/de/aktuelles_und_presse/pressemitteilungen/detail/item/the_federal_minister_of_education_and_research_inaugurates_new_german_antarctic_station_neumayer_s/?cHash=561ffe6b41e4313caa2835af2cdba502 (accessed June 3, 2015).

AWI weekly report ANTXXIV/3 Wochenbericht No. 3 from Sunday, March 9, 2008: https://www.awi.de/de/infrastruktur/schiffe/polarstern/wochenberichte/alle_expeditionen/ant_xxiv/ant_xxiv3/9_maerz_2008/ (accessed June 3, 2015).

CryoSat: www.esa.int/Our_Activities/Observing_the_Earth/CryoSat/Introducing_CryoSat (accessed November 27, 2019); www.esa.int/ger/ESA_in_your_country/Germany/Drei_Jahre_CryoSat_Das_neue_Bild_vom_irdischen_Eis (accessed November 27, 2019).

Dallmann Laboratory: https://www.awi.de/en/expedition/stations/dallmann-laboratory.html (accessed November 27, 2019).

Drescher Station: https://www.awi.de/en/expedition/stations/drescher-ice-camp.html (accessed November 27, 2019).

Filchner Station: https://de.wikipedia.org/wiki/Filchner-Station (accessed November 27, 2019).

Filchner Station retrieval: https://antarktis.ch/2000/01/01/antarktis-xvi-1998-99/ (accessed November 27, 2019).

Gondwana Station: https://www.bgr.bund.de/EN/Themen/Polarforschung/Antarktis/Logistik/logistik_node_en.html (accessed November 27, 2019).

Gotland II sinking: http://vphn-os.de/wp-content/uploads/2015/12/Auszug-aus-VN4-2015_Weihnachten-1981-Untergang-der-Gotland-II.pdf (accessed July 17, 2020).

Historic Sites and Monuments in Antarctica: https://documents.ats.aq/documents/ATCM36/WW/atcm36_ww004_e.pdf (accessed November 27, 2019).

History of Antarctic Research: https://www.scar.org/science/hass/history-group/ (accessed November 27, 2019).

History of Polar Research: https://www.polarforschung.de/arbeitskreise/ak-geschichte-der-polarforschung/ (accessed November 26, 2019).

Horizon Scan: https://www.scar.org/science/horizon-scan/overview/ (accessed December 3, 2020); www.uc.pt/tomenota/2014/082014/documentos/201 40806_1 (accessed November 27, 2019).

IceCube: http://icecube.wisc.edu/ (accessed November 27, 2019).

ICESat: http://icesat.gsfc.nasa.gov/icesat/ (accessed November 27, 2019).

Kohnen Station: https://www.awi.de/en/expedition/stations/kohnen-station.html (accessed November 27, 2019).

Neumayer III Station: https://www.awi.de/en/expedition/stations/neumayer-station-iii.html (accessed November 27, 2019).

O'HigginsStation: https://www.dlr.de/eoc/en/desktopdefault.aspx/tabid-9472/16238_read-40703/ (accessed November 26, 2019).

»Polarstern«: https://www.awi.de/en/expedition/ships/polarstern.html (accessed November 27, 2015).

»Polarstern« (Cruise Report): http://epic.awi.de/26234/1/BerPolarforsch199057.pdf (accessed February 7, 2015); http://epic.awi.de/26447/1/BerPolarforsch1998267.pdf (accessed November 27, 2019); https://antarktis.ch/2000/01/01/antarktis-xvi-1998-99/ (accessed November 27, 2019); https://www.awi.de/de/infrastruktur/schiffe/polarstern/wochenberichte/alle_expeditionen/ant_xxiv/ant_xxiv3/9_maerz_2008/ (accessed June 6, 2015).

Press release about the newest icebreaker: https://www.awi.de/de/aktuelles_und_presse/pressemitteilungen/detail/item/milestone/?cHash=6dbb504586108c76 8d28f0e7ccaf6d8d (accessed June 3, 2015).

Princess Elisabeth Antarctica Station: www.antarcicstation.org (accessed July 17, 2020).

Split of iceberg A38 in October 1998: https://de.wikipedia.org/wiki/A-38; https://en.wikipedia.org/wiki/Iceberg_A-38 (accessed November 27, 2019).

Protocol on Environmental Protection (Madrid Protocol): https://www.coolantarctica.com/Antarctica%20fact%20file/science/madrid_protocol.php (accessed November 27, 2019).

Wallnerspitze: www.data.antarctica.gov.au/aadc/gaz/display_name.cfm?gaz_id=107149; https://de.wikipedia.org/wiki/Wallnerspitze (accessed November 27, 2019).

Weigelnunatak: https://data.aad.gov.au/aadc/gaz/scar/display_name.cfm?gaz_id=107152; https://de.wikipedia.org/wiki/Weigelnunatak (accessed November 27, 2019).

Picture Credits

Albrecht, Klaus-Peter (Private possession): Figs. 4.7, 4.9, 4.10, 4.11.
Alfred-Wegener-Institut/Thomas Steuer, licensed under the terms of CC-BY 4.0 (https://creativecommons.org/licenses/by/4.0/legalcode): Figs. 4.38, 4.39, 4.40 (detail), 4.41, 4.42.
Archiv für deutsche Polarforschung, Alfred-Wegener-Institut, Bremerhaven, Grotewahl estate: Figs. 2.43, 2.44.
Bergman, Paul (Private possession): Fig. 4.2.
Bundesanstalt für Geowissenschaften und Rohstoffe, Hanover: Fig. 4.6.
Bundesamt für Seeschifffahrt und Hydrographie, Hamburg: Fig. 1.1.
Deutsches Schiffahrtsmuseum, Bremerhaven: Fig. 3.20.
Die Weite Welt No. 28 issue of July 9, 1939: Fig. 3.18.
Drygalski, Erich von: Zum Kontinent des eisigen Südens, Berlin 1904, p. 81: Fig. 1.4, pp. 64–65: Fig. 1.8, pp. 254–255: Fig. 1.61.
Drygalski, Erich von (ed.): Deutsche Südpolar-Expedition 1901–1903. Berlin, 1905–1931: Vol. 1, Table 1: Fig. 1.42, Vol. 7, Table 3: Fig. 1.20.
Enß, Dietrich: Die deutsche Antarktis-Forschungsstation, in: Hansa 118 (1981) 13, p. 965: Fig. 4.8.
Fahrbach, Karin (Private possession): Figs. 4.29, 4.30, 4.31.
Filchner Archive, Munich: Figs. 2.1, 2.3, 2.9, 2.10, 2.11, 2.12, 2.16, 2.42.
Filchner, Wilhelm: Plan einer deutschen antarktischen Expedition, in: Zeitschrift der Gesellschaft für Erdkunde zu Berlin 45 (1910) 3, p. 154: Fig. 2.2.
Filchner, Wilhelm: Zum sechsten Erdteil. Berlin 1922, p. 27: Fig. 2.5, p. 32: Fig. 2.4, p. 42: Fig. 2.14, p. 190: Fig. 2.28, p. 198: Fig. 2.32, p. 211: Fig. 2.29, p. 321: Fig. 2.34; p. 306: Fig. 2.38, map (detail): Fig. 2.26.
Filchner, Wilhelm: Ein Forscherleben, Wiesbaden 1950, p. 97: Fig. 2.6.
Filchner, Wilhelm/Seelheim, Heinrich: Quer durch Spitzbergen. Berlin 1911, Tf 11: Fig. 2.7.
Förster, Hans Albert: Der Hohe Pol. Leipzig 1956, p. 203: 1.12.
Gazert, Volker (Private possession): Figs. 1.3, 1.10, 1,13, 1.16, 1.17, 1.26, 1.28, 1.43, 1.44, 1.46, 1.50, 1.51, 1.59, 1.65.

Geographisches Institut der Universität, Munich: Fig. 1.55.
Gernandt, Hartwig (Private possession): Fig. 4.28.
Gernandt, Hartwig; Erlebnis Antarktis, Berlin 1984, pp. 258–259: Fig. 4.3.
Gravenhort, Gode: Antarktis-Expedition 1981/1982 (Unternehmen »Eiswarte«), in: Berichte zur Polarforschung 6/'82 (1982), p. 17: 4.13.
Hartmann, Gertraude (Private possession): Figs. 3.46, 3.47.
Herrligkoffer, Karl Maria: Deutsche Südpol-Expedition, Munich 1956, p. 4: Fig. 4.1.
Herrmann, Ernst: Deutsche Forscher im Südpolarmeer, Berlin 1941, p. 61: Fig. 3.26, p. 429: Fig. 3.48.
Joester, Erich (Private possession): Figs. 2.13, 2.15, 2.17, 2.18, 2.19, 2.20, 2.21, 2.23, 2.24, 2.25, 2.27, 2.30, 2.31, 2.33, 2.35, 2.36, 2.37, 2.39, 2.40, 2.41, 3.50, 5.1.
Kennicutt, Chuck (Private possession): Fig. 5.2.
Kleinschmidt, Georg (Private possession): Figs. 4.12, 4.19. 4.20.
Leibniz-Institut für Länderkunde, Leipzig, Drygalski estate: front matter, 1.2, 1.6, 1.9, 1.14, 1.19, 1.22, 1.23, 1.24, 1.27, 1.31, 1.37, 1.39, 1.47, 1.54.
Leibniz-Institut für Länderkunde, Leipzig, Herrmann estate: Figs. 3.9, 3.27, 3.28, 3.30, 3.36, 3.40, 3.41, 3.42.
Leibniz-Institut für Länderkunde, Leipzig, Ritscher estate: Figs. 3.6, 3.8, 3.12, 3.13, 3.14, 3.15, 3.16, 3.19, 3.22, 3.23, 3.24, 3.25, 3.29, 3.31, 3.32, 3.33, 3.34, 3.35, 3.37, 3.38, 3.39, 3.43, 3.44, 3.45.
Lüdecke, Cornelia (Private possession): Figs. 1.19, 1.29, 1.30, 1.32, 1.66, 2.8, 2.22, 3.49, 4.17, back matter
Lüdecke, Cornelia: Die erste deutsche Südpolar-Expedition und die Flottenpolitik unter Kaiser Wilhelm II, in: Historisch-meereskundliches Jahrbuch 1 (1992), p. 64: Fig. 1.11.
Mörder, Thomas (Private possession): Figs. 1.21, 1.25, 1.33, 1.34, 1.35, 1.36, 1.38, 1.40, 1.41, 1.45, 1.48, 1.49, 1.52, 1.53, 1.56, 1.57, 1.58, 1.60, 1.62, 1.63, 1.64.
Oberhummer, Eugen: Die Deutsche Südpolarexpedition. 18. Jahresbericht der Geographischen Gesellschaft in München 1900, Table V: Fig. 1.5, Table II: Fig. 1.7.
Oerter, Hans (Private possession): Figs. 4.14, 4.18, 4.32, 4.33, 4.34, 4.35, 4.36, 4.37, 5.3.
Peters, Nicolaus: Über Hochseewalfang und Tierleben im Südlichen Eismeer, in: Der Fischmarkt (1937) 7/8, p. 2: Fig. 3.5.
Puskeppeleit, Monika (Private possession): Figs. 4.21, 4.22, 4.24, 4.25, 4.26.
Regula, Herbert: Die Wetterverhältnisse während der Expedition und die Ergebnisse der meteorologischen Messungen, in: Ritscher, Alfred (ed.): Deutsche Antarktische Expedition 1938/39. Wissenschaftliche Ergebnisse Vol. 2, Issue 1, Hamburg 1954, p. 25: Fig. 3.51.
Ritscher, Alfred (ed.): Wissenschaftliche und fliegerische Ergebnisse der Deutschen Antarktischen Expedition 1938/39, Leipzig 1942, p. 5: Fig. 3.7, after p. 16, Fig. 3.10, p. 23: Fig. 3.11, p. 29: Fig. 3.21.
Schönhofer, Georg (Private possession): Fig. 4.15.
Schug, Joachim (Private possession): Fig. 4.16.

Spengemann, Herbert: Auf Walfang in der Antarktis, Bühl-Baden 1939, p. 65: Fig. 3.1, p. 16: Fig. 3.2, p. 100: Fig. 3.4, p. 103: Fig. 3.3.
Strecke, Volker (Private possession): Figs. 4.4, 4.5, 4.23, 4.27.
Stuttgarter Illustrierte No. 37, issue of 10 September 1939: Fig. 3.17.
Terra Marique 1902, Issue 1: Fig. 1.15.

Abbreviations

AWI	Alfred-Wegener-Institut für Polarforschung
BGR	Bundesanstalt für Geowissenschaften und Rohstoffe
BRD	Bundesrepublik Deutschland
DAE	Deutsche Antarktische Expedition
DDR	Deutsche Demokratischen Republik
EPICA	European Project for Ice Coring in Antarctica
ERS	European Remote Sensing
ESA	European Space Agency
FRG	Federal Republic of Germany
GANOVEX	German Antarctic North Victoria Land Expedition
GARS	German Antarctic Receiving Station
GDR	German Democratic Republic
ICESat	Ice, Cloud, and Land Elevation Satellite
NSIU	Norges Svalbard- og Ishavsundersøkelser (today: Norsk Polarinstititt)
PALAOA	Perennial Acoustic Observatory in the Antarctic Ocean
SAE	Soviet Antarctic Expedition
SCAR	Scientific Committee on Antarctic Research
ZIPE	Zentralinstitut für Physik der Erde

Chronology of the History of German Antarctic Research

July 24, 1865	First Convention of the Masters and Friends of Geography [Erste Versammlung Deutscher Meister und Freunde der Erdkunde] in Frankfurt; Georg von Neumayer's unsuccessful attempt to dispatch a German South Polar expedition
1873–1874	First German expedition to the Antarctic Peninsula, led by Eduard Dallmann on the steamship *Groenland*: exploration of new whaling grounds
1874–1875	German expedition to the Kerguelen Islands: observation of the transit of Venus on December 9, 1874
August 1, 1882–September 1, 1883	First International Polar Year in the Arctic; additional German meteorological and magnetic station in South Georgia and observation of the transit of Venus on December 6, 1882
May 1–September 18, 1891	Erich von Drygalski's pre-expedition to the western coast of Greenland near Umanak
May 1, 1892–October 14, 1893	Erich von Drygalski's main expedition to the western coast of Greenland: overwintering in a station at the Small Karajak Ice Stream near Umanak
June 24, 1893–September 9, 1896	Expedition led by the Norwegian Fridtjof Nansen on board the *Fram*: an attempt to reach the North Pole
April 17–19, 1895	11th Conference of German Geographers in Bremen: foundation of the German Commission for South Polar Research [Deutsche Kommission für Südpolarforschung], directed by Georg von Neumayer
July 26–August 3, 1895	Sixth International Geographical Congress in London: resolution to commence Antarctic research
February 19, 1898	Erich von Drygalski appointed as leader of the first German South Polar Expedition
September 28–October 4, 1899	Seventh International Geographical Congress in Berlin: agreement on international meteorological and magnetic cooperation in Antarctica
April 2, 1901	Launch of the first German polar research vessel, the *Gauss*, at the Howaldt Shipbuilding Yard in Kiel
May 28, 1901	Thirteenth Conference of the German Geographers in Breslau: dissolution of the German Commission for South Polar Research
August 11, 1901	Departure of the first German South Polar Expedition, led by Erich von Drygalski on the *Gauss*, from Kiel

(continued)

October 1, 1901–March 31, 1903	International meteorological and magnetic cooperation in Antarctica, with participation by the German expedition led by Erich von Drygalski on the *Gauss*, the British expedition led by Robert Falcon Scott on the *Discovery*, the Swedish expedition led by Otto Nordenskjöld on the *Antarctic*, and the Scottish expedition led by William Spears Bruce on the *Scotia*; extended for one year until March 31, 1904
November 25, 1903	Return of the German South Polar Expedition to Kiel: discovery of Gaussberg in Kaiser Wilhelm II Land at 90°E
March 15, 1904	Sale of the *Gauss* to the Canadian government; deployment under the name *Arctic* for surveys in the Canadian Arctic region
March 1910	Meeting of the Berlin Geographical Society [Gesellschaft für Erdkunde zu Berlin]: Wilhelm Filchner's presentation of an expedition plan to explore the relationship between East and West Antarctica
August 4–18, 1910	Wilhelm Filchner's training expedition to Spitsbergen: traverse of the unexplored region between Isfjorden and Storfjorden
January 3, 1911	Constituent assembly of the Committee for the German Antarctic Expedition [Komitee für die Deutsche Antarktische Expedition] in Berlin
1910–1914	Race to the South Pole, including Robert Falcon Scott (leader of the British expedition on the *Terra Nova*), Roald Amundsen (leader of the Norwegian expedition on the *Fram*), and Nobu Shirase (leader of the Japanese expedition on the *Kainan Maru*), besides Wilhelm Filchner's German expedition on the *Deutschland*
February 19–April 20, 1911	Reconstruction of the *Deutschland* for the Antarctic expedition at the shipyard of Blohm & Voß in Hamburg
May 3, 1911	Departure of the second German Antarctic Expedition [Deutsche Antarktische Expedition], led by the officer and topographer Wilhelm Filchner, on the *Deutschland* from Hamburg
November 26, 1911	Suicide of the third officer of the *Deutschland*, Walter Slossarczyk, in South Georgia
December 14, 1911	Roald Amundsen is the first person to reach the South Pole
August 6, 1912	Departure of Herbert Schröder-Stranz's training expedition to Spitsbergen on the *Herzog Ernst* (captain: Alfred Ritscher) from Tromsø
August 8, 1912	Death of Richard Vahsel, the captain of the *Deutschland*
December 19, 1912	Official end of the second German Antarctic Expedition in Grytviken in South Georgia after a mutiny; discovery of Prinzregent Luitpold Land (in the eastern part of the Weddell Sea)
December 27, 1912	Ritscher's arrival at the coal settlement Advent Bay: beginning of the relief campaign for the members of the Schröder-Stranz expedition overwintering at various sites and those who were lost
1914–1918	World War I
July 1, 1926	Foundation of the Archive for Polar Research [Archiv für Polarforschung] in Kiel under the directorate of Max Grotewahl
1927	Foundation of the Association for the Promotion of the Polar Research Archive [Vereinigung zur Förderung des Archivs für Polarforschung]
1931	Publication of the results of Drygalski's German South Polar Expedition in 20 volumes and two atlases
1931	Release of the journal Polarforschung [Polar Research]
August 1, 1932–August 31, 1933	Second International Polar Year in the Arctic, without observation stations in Antarctica

(continued)

Chronology of the History of German Antarctic Research

November 13, 1936	First departmental discussion at the German Foreign Office: protection of German whaling activities by territorial claims in Antarctica
July 1937	Discovery of the remains of the lost Schröder-Stranz expedition in Nordaustlandet (North East Land), Spitsbergen
May 9, 1938	Submission of a plan for a third German Antarctic expedition to Reich Marschal Hermann Göring by Helmuth Wohlthat
July–August 1938	Ernst Herrmann's expedition to Spitsbergen: deployment of a Fieseler-Storch Langsamflugzeug (slow-flight aircraft)
September 1, 1938	Appointment of the government official, captain, and aircraft pilot Alfred Ritscher as the leader of the third German Antarctic Expedition by the Four-Year Plan (Field Marshal Göring)
November 4–6, 1938	Preliminary tests for drift ice research: testing of ice arrows on the Pasterzen Glacier (Austria) by Ernst Herrmann
December 17, 1938	Departure of the third German Antarctic Expedition (DAE 38/39) from Hamburg, led by Alfred Ritscher on the catapult ship *Schwabenland*
April 11, 1939	Return of the third German Antarctic Expedition (DAE 38/39) to Hamburg: discovery and exploration of Neu-Schwabenland with the aircraft *Boreas* and *Passat*
June 1939	Ritscher's draft program for the subsequent fourth German Antarctic Expedition (DAE 39/40)
September 1, 1939	Sudden discontinuation of preparation for DAE 39/40 due to the outbreak of World War II
1939–1945	World War II
1951	First International Polar Conference of the Association for the Promotion of the Archive for Polar Research in Kiel: Alfred Ritscher elected as the first chairman
1956–1958	Karl-Maria Herrligkoffer's plan for German participation in the International Geophysical Year with a station in Neu-Schwabenland
July 1, 1957– December 31, 1958	International Geophysical Year (subsequently referred to as the Third International Polar Year); first research stations in Antarctica manned all year round
January 1– December 31, 1959	Extension of the International Geophysical Year for one year as the International Geophysical Cooperation
1959	Second International Polar Conference of the Association for the Promotion of the Archive for Polar Research in Holzminden: reorganization of the association, then called German Society for Polar Research [Deutsche Gesellschaft für Polarforschung e.V.]
1959	Organization of Antarctic research in the German Democratic Republic (GDR) by the National Committee for Geodesy and Geophysics [Nationalkomitee für Geodäsie und Geophysik]
1960	Overwintering of the first GDR research group at the Soviet Mirny Station in Antarctica; start of continuous participation of GDR scientists in Soviet Antarctic expeditions
1969	Transfer of the organization of Antarctic research in the GDR to the Central Institute of Physics of the Earth [Zentralinstitut für Physik der Erde] in Potsdam
November 19, 1974	Accession of the GDR to the Antarctic Treaty
1976	Establishment of the GDR base station close to the Soviet Novolazarevskaya Station in Schirmacher Oasis

(continued)

May 22, 1978	Acceptance of the Federal Republic of Germany (FRG) as a member of the Scientific Committee on Antarctic Research (SCAR)
February 5, 1979	Accession of the FRG to the Antarctic Treaty
1979–1980	Exploratory expedition to seek a site for the FRG Antarctic station on the Filchner–Ronne Ice Shelf
1979–1980	Beginning of Antarctic research by the Federal Institute for Geosciences and Natural Resources [Bundesanstalt für Geowissenschaften und Rohstoffe (BGR)]: dispatch of the first expedition (GANOVEX I) to North Victoria Land (East Antarctica)
January 14, 1980	Establishment of the Lillie Marleen Hut in North Victoria Land by the BGR
July 15, 1980	Foundation of the Alfred Wegener Institute for Polar Research [Alfred-Wegener-Institut für Polar Forschung (AWI)] in Bremerhaven
February 24, 1981	Roofing ceremony at the FRG's Georg von Neumayer Station on the Ekström Ice Shelf near Atka Bay
March 3, 1981	Accession of the FRG to the Antarctic Treaty as a member state with consultative status
September 9, 1981	Acceptance of the GDR as a member of SCAR
December 18–19, 1981	Sinking of the *Gotland II* with the expedition GANOVEX II from the BGR on board; evacuation of all persons to the Lillie Marleen Hut
January 11, 1982	Inauguration of the FRG's Filchner Station for summer operations on the Filchner–Ronne Ice Shelf
January 25, 1982	Christening of the FRG's polar research icebreaker *Polarstern*
January 1983	Establishment of the BGR's Gondwana Hut in North Victoria Land
1983–1984	First deployment of the polar aircraft *Polar 1* (Do 128) and *Polar 2* (Do 228) at Georg von Neumayer Station
1984	Sale of *Polar 1* to the Technical University of Brunswick in Germany
December 1984	Flight of the *Polar 2* and the chartered *Polar 3* from Georg von Neumayer Station over the South Pole to McMurdo on the Ross Sea; survey flights from the BGR's Gondwana Camp
February 24, 1985	On its return flight from Georg von Neumayer Station to Germany, *Polar 3* is shot down over the Western Sahara coast by Polisario rebels
End of 1985	Commissioning of the polar aircraft *Polar 4*
October 1986	Establishment of Drescher Station by the AWI as a mobile biologist camp on the Riiser-Larsen Ice Shelf on the northeastern coast of the Weddell Sea
October 25, 1987	Renaming of the GDR's research base in the Schirmacher Oasis as Georg Forster Station; accession of the GDR to the Antarctic Treaty as a member state with consultative status
1990	First and globally unique Antarctic overwintering by an all-female team at Georg Neumayer Station; introduction of complete waste separation, composting of organic waste, return of all refuse to Germany, and use of wind power turbines
October 3, 1990	Signing of the German Reunification Treaty
January 10, 1991	Opening of the German Antarctic Receiving Station (GARS) O'Higgins on the northern point of the Antarctic Peninsula near the Chilean Station Bernardo O'Higgins by the German Aerospace Center [Deutsches Zentrum für Luft und Raumfahrt]; reception of measurement data from the European remote sensing satellite ERS-1

(continued)

Chronology of the History of German Antarctic Research

October 4, 1991	Signing of the Protocol on Environmental Protection by the Antarctic Treaty members in Madrid
December 1991	Replacement of the last GDR overwinterers at Georg Forster Station by the *Polarstern*
March 11, 1992	Inauguration of the research department at the Alfred Wegener Institute on Telegrafenberg in Potsdam
March 31, 1992	Opening of Neumayer Station II on the Eckström Ice Shelf
February 1993	Closing of the GDR's Georg Forster Station in East Antarctica
January 1994	Installation of the Dallmann Laboratory by the AWI and the Argentinian Antarctic Institute for summer operations near the Argentinian Carlini Station (formerly Jubany Station) on King George Island
1995–1996	Extension of investigations by the BGR in Dronning Maud Land (East Antarctica)
1996	Complete removal and disposal of the GDR's Georg Forster Station
October 13, 1998	Ice island A-38 breaks off from the Filchner–Ronne Ice Shelf on which Filchner Station stands
February 1999	Complete retrieval of the Filchner Station
Summer 2000–2001	Establishment of Kohnen Station at an altitude of 2892 meters and 750 kilometers away from Neumayer Station II for the European Project for Ice Coring in Antarctica (EPICA)
2002–2003	Extension of BGR research to central East Antarctica (Lambert Glacier, Gamburtsev Mountain Range)
January 2005	Decommissioning of *Polar 4* after a hard landing on ice
January 2006	Finalization of the deep ice coring at Kohnen Station
October 1, 2007	Commissioning of *Polar 5*, an adapted Basler BT-67 built in 1942
March 1, 2007– March 1, 2008	Fourth International Polar Year, already extended for one year (until March 1, 2009) during planning
February 20, 2009	Inauguration of Neumayer Station III on the Eckström Ice Shelf
April 8, 2010	Launch of the CryoSat-2 satellite to explore earth's cryosphere
October 28, 2011	Commissioning of *Polar 6*, an adapted Basler BT-67 built in 1943
April 20–23, 2014	So-called Horizon Scan conducted by the SCAR in Queenstown (New Zealand): elaboration of the 80 most important questions for future research projects in Antarctica

About the Author

Fig. 1 Cornelia Lüdecke. (Lüdecke, Cornelia (Private possession))

Cornelia Lüdecke, Prof. Dr. rer. nat., is a meteorologist and a science historian. She has been the head of the German Society for Polar Research's History of Polar Research Working Group since 1991. From 1995 to 2018, she was the head of the German Meteorological Society's Expert Committee on the History of Meteorology; the Society awarded her the Reinhard Süring Medal in 2010 for her achievements and the Paulus Price for the History of Meteorology in 2019. From 2001 to 2005, she was the vice-president and from 2006 to 2009 the president of the International Commission on the History of Meteorology. She has been the chairwoman of the Scientific Committee on Antarctic Research's History of Antarctic Research Expert Group since 2004, and she has been a corresponding member of the International

Academy of the History of Science in Paris since 2012. She has published 20 monographs and more than 180 articles on the history of meteorology, geography, oceanography, and polar research, and she has been a scientific advisor to the German television organizations ARD, Arte, Bayerischer Rundfunk (BR), Spiegel TV, and Westdeutscher Rundfunk (WDR), among others. She herself has also participated in numerous field studies and trips to polar regions (Fig. 1).

Geographical Index

A
Advent Bay, 86
Advent City, 142
Africa, 68, 71, 195, 226
Alaska, 202
Alexander Land, 25
Amundsen-Scott Station, 212
Antarctic Peninsula (formerly Graham Land), 4, 6, 22, 79, 220, 222, 235
Arctic, viii, 3, 5, 9, 11, 25, 74, 128, 134, 152, 153, 162, 163
Arctic Ocean, 5, 25, 152, 232
Argentina, 22, 97–99, 127, 143, 187, 188, 195, 222
Ascension, 71, 157
Atka Bay, 202–204
Atlantic Ocean, 20, 27, 32, 34, 35, 93, 98, 128, 145, 147, 148, 152–154, 157, 164, 184, 185, 189
Austria, 131, 152
Azores, 71

B
Bad Lauterberg, 141, 145
Bavarian Forest, 92, 93
Bay of Biscay, 144
Belgium, 195
Berlin, 1–9, 11, 13, 17, 18, 20–22, 28, 42, 48, 66, 68, 69, 73, 76–79, 81, 83, 86–91, 93, 96, 104, 116, 125–129, 134, 136, 139, 141, 145, 152, 160, 162, 163, 198, 200, 206, 208, 218, 230
Bernardo O'Higgins Station, 220
Bertrab Nunatak, 109
Bismarck Strait, 4

Bonn, 7, 201
Bouvet Island, 153, 157, 180
Bremen, 5, 98
Bremerhaven, 93–96
Breslau, 6, 11, 26
British India, 78
Buenos Aires, 83, 89, 93, 95–98, 100–103, 105, 126, 143, 187

C
Camp Capsize, 215
Canary Islands, 155
Cape Adare, 11
Cape Finisterre, 155
Cape Town, 32, 37, 38, 44, 68, 69, 163, 180, 226, 229
Carlini Station (formerly Jubany Station), 222
Central Asia, 76, 78, 196–197
Chile, 195, 212, 235
China, 1, 2, 76, 129
Christiana (now Oslo), 24, 152, 162
Coats Land, 22, 90, 161
Cochoeiro Bay, 181
Conrad Mountains, 178, 181
Crozet Islands, 39
Cumberland Bay, 99
Cuxhaven, 182

D
Drescher Station, 213
Dronning Maud Land (formerly Queen Maud Land), 162, 163, 187, 200, 225, 227, 228, 235
Drygalski Fjord, 102

Drygalski Island, 197
Drygalski Mountains, 173, 177
Durban, 68

E
East Europe, 221
East Tibet, 76
Egersund, 187
Ekström Ice Shelf, 202, 210
Enderby Land, 25
England, 2, 75, 80, 92, 139
English Channel, 144, 182, 187
Europe, 78, 162, 221

F
Falkland Islands, 22
Farther India, 1
Federal Republic of Germany (FRG), vii, 193–195, 200–216, 218, 221, 233
Fernando Noronha, 182
Filchner Ice Shelf, 108
Filchner Mountains, 177
Filchner Ronne Ice Shelf, 202, 208, 223
Filchner Station, 208, 209, 223, 225
Fortuna Bay, 102
France, 75, 139, 195
Frankfurt-on-Main, 143
Franz Joseph Fjord, 3
Franz Joseph Land, 6
Friedrichshafen, 213

G
Gaussberg, 50, 52, 53, 56, 58, 59
Georg Forster Station, 198, 200, 201, 217, 218, 221–223
Georg von Neumayer Station, 200, 203–213, 216–219
German Democratic Republic, 196–201, 223
German Samoa, 22
Germany, vii, viii, 1–12, 20–22, 58, 62, 73, 75, 78, 85, 126, 131, 133, 134, 136, 139, 145, 153, 162, 163, 180, 183, 187, 193–195, 200–215, 218, 221, 229, 232, 233, 235
Gondwana (continent), 202, 209, 212, 237
Gondwana Hut, 209
Gondwana Station, 212
Gottingen, 11, 22, 166, 186
Graham and Alexander Land, 25

Gralsburg, *see* Filchner Mountains
Grautskåla Cirque, 179
Great Britain, 20, 73, 75, 78, 195, 235
Greenland, 3, 6–8, 25, 60, 85, 92, 122, 131, 152
Grossglockner, 152
Grottenberg, v
Grytviken, 83, 101–103, 105, 117, 124–127

H
Hamburg, 4, 5, 93, 101, 103, 134, 136, 141, 145, 155, 160, 164, 173, 182–184, 189, 190, 195, 202, 206
Heard Island, 42, 43
Helsinki, 218
Herzog Ernst Bucht, *see* Vahsel Bay
Holland, 5, 78
Humboldt Mountain Range, 179, 184

I
Indian Ocean, 4, 8, 22, 68, 128
Isfjorden, 83

J
Japan, 1, 195
Jena, 5, 148

K
Kaiser Franz Joseph Fjord, 3
Kaiser Wilhelm Canal, 32, 71
Kaiser Wilhelm II Land, 8, 21, 43, 73, 125, 187
Kaiser Wilhelm Islands, 4
Kamchatka, 37
Karajak Ice Stream, 269
Karlsruhe, 37, 213
Kavrayskiy Hills, 215
Kemp Land, 25
Kerguelen Station, 22, 41, 66, 69
Kiel, 17, 20, 24, 25, 27, 32, 33, 71, 130, 131, 208
Kiel Canal, 32
King Edward Cove, 104
King George Island, 222
King Haakon Bay, 102
Knox Land, 62
Kohnen Station, 225, 227, 229, 230
Korcula, 128
Kottas Camp, 224, 226, 227

Geographical Index 279

L
Lake Titicaca, 60–62, 64
Leipzig, vii, 1, 7, 8, 11, 12, 15–17, 19, 23–26, 28–30, 35, 36, 39, 41, 47, 49, 55, 61, 70, 76, 133, 138–141, 145, 148, 153, 154, 156, 160, 162–164, 166, 167, 181, 182, 185–187, 197
Lillie Glacier, 202
Lillie Marleen Hut, 202, 203, 207
London, 4, 5, 79, 139, 184, 186, 188
Lower Elbe, 32
Luitpold Coast, 108

M
Madrid, 221, 273
Mar del Plata, 187
Martin Vaz (Brazilian: Martim Vaz), 152, 153, 181, 186
McMurdo Station, 212
Mid-Atlantic Ridge (formerly Atlantic Threshold), 34, 93, 184, 185
Mirny Station, 197
Mohn Bay, 83
Molodezhnaya Station, 198
Munich, 1, 8, 9, 11, 66, 76–78, 87, 89, 90, 103, 127, 165, 193, 195, 200, 201

N
Namibia, 70
Natal, 143
Nepal, 78
Neumayer Channel, 4
New Amsterdam (island), 68
New Guinea, 1
New Zealand, viii, 20, 22, 195, 214, 236, 237
North Italy, 144
North pole, 6, 25, 162
North Victoria Land, 202, 203, 208, 209
Norway, 6, 24, 78, 81, 85, 86, 134, 139, 152, 157, 162, 163, 195
Novolazarevskaya Station, 198, 199, 226

O
Oates Land, 215
Oldenburg, 133, 143

P
Paulet Island, 101
Peary Land Peninsula, 152
Pernambuco, 96, 190
Pilar, 98
Poona, 78
Posadowsky Bay, 58
Possession Island, 39
Potsdam, 25, 76, 197, 221
Prince Luitpold Glacier (now Tunabreen), 86
Princess Elisabeth Antarctica Station, 235
Principe Bay, 181
Prinzregent-Luitpold-Land, 108, 110, 111, 115
  See also Luitpold Coast
Punta Arenas, 212

Q
Queen Maud Land (now Dronning Maud Land), 235
Queenstown, 236, 237

R
Recife, 180–182
Riiser-Larsen Ice Shelf, 213
Rijeka, 128
Rio de Janeiro, 143, 152
Ritscher Upland, 213, 214
Romanche Deep, 34, 70
Ross Sea, 44, 61, 79, 80, 86, 127, 128, 145, 195, 202, 207, 212
Royal Bay, 99, 102
Russia, 7, 78, 86, 197, 201, 221, 226

S
Sandwich Bay (Gold Harbor), 102
Sandwich Islands, 100, 101, 103
Schirmacher Oasis (formerly Boreasic Lake District), 173, 179, 198–200, 226
Shetland Islands, 222
Siberian coast, 25
Simon's Town, 68, 71
South Africa, 68, 71, 195, 226
South America, 96, 139–141, 143, 164, 189, 190
South Georgia, 83, 96, 98–105, 114, 117, 124, 126, 236
South Patagonia, 97
South Pole, vii, 4, 14, 25, 26, 28, 67, 73, 78, 80, 81, 87, 100, 105, 109, 195, 212, 235–237
South Sea, 1, 91
Soviet Union, 195, 200

Spitsbergen, 3, 77, 83–87, 129, 130, 140, 141, 152, 156, 163
Staten Island, 22
St. Helena, 71
Stockholm, 85
Storfjorden, 83
St. Paul, 68, 96
Stromness Bay, 103
Switzerland, viii, 131
Sydney, 37

T
Temple Bay, 83
Tenerife, 182
Tian Shan Mountains, 196
Tibet, 76, 78, 129, 159
Tierra del Fuego, 99
Treurenburg Bay, 141
Trinidade, 152, 153, 181, 186
Tristan da Cunha, 157
Troll Station, 229
Tromsø, 85
Trypot Bay (formerly Boiler Bay), *see* King Edward Cove
Tsingtau (today: Qingdao, Shandon province, China), 2
Tunabreen, 86

U
Umanak, 269

United States of America, 139, 189, 195, 196, 239

V
Vahsel Bay, 109–111, 113–115, 120
Verdun, 78
Victoria Land, 11, 25, 202, 203, 208, 209, 213, 215
Vienna, 11

W
Walvis Ridge, 70
Washington, D.C., 196
Weddell Sea, 14, 20, 22, 25, 77, 79, 86, 87, 96, 101, 114, 116, 123, 127, 128, 195, 202–209, 213
Wellington, 214
West Sahara coast, 212
Wilhelmshaven, 127
Wilkes Land, 25
Wohlthat Massif, 173, 184, 200
    See also Wohlthat Mountains
Wohlthat Mountains, 173
    See also Wohlthat Massif
Wrangel Island, 25

Z
Zavodovski Island, 103, 104
Zurich, 78

Person Index

A
Adler, Wieland, 221
Albert II, von Monaco, 196
Amundsen, Roald (1872–1912), 128, 129, 270
Ankersen, Walter, 130

B
Bähr, Gustav (born. 1877), 22
Balch, Edwin Swift (1856–1927), 79
Barkley, Erich (1912–1945), 170, 173, 177, 181
Barkow, Erich (1882–1923), 85–87, 90, 98–101, 103, 105–107, 115–122, 124–128, 241
Barnes, Jim, 196
Baudissin, Graf Friedrich von (1852–1921), 10, 11, 15
Baumgart, Susanne, 216
Berglöf, Emil (born 1879), 19
Bergman, Paul Arthur, 196
Berguño, Jorge (1929–2011), 196
Bernier, Joseph-Elzéar (1852–1934), 74
Bertrab, Hermann von (1857–1940), 77, 79, 91
Besenbrock, August (born 1882), 19, 83, 92
Bethmann Hollweg, Theobald von (1856–1921), 75, 76
Beyer, Reinhard, 211
Bidlingmaier, Friedrich (1875–1914), 10, 22, 25, 27, 32, 34, 37, 41, 44, 46, 54–56, 58
Biller, Fritz, 130
Björvik, Paul (born 1857), 19, 81, 83, 92, 96, 98, 102, 103, 105, 108, 111, 112, 120, 125
Bludau, Josef (born 1889), 149, 182
Borchgrevink, Carsten (1864–1934), 11
Böttcher, Fritz, 83, 111
Brennecke, Wilhelm (1875–1924), 88, 93, 102, 103, 114, 116–119, 121, 125, 127, 128
Bruce, William Speirs (1867–1921), 22, 23, 79–81, 86, 270
Bruns, Herbert (born 1908), 170, 193
Bundermann, Max (born 1904), 152, 161
Byrd, Richard Evelyn (1888–1957), 188, 189, 193

C
Charcot, Jean-Baptiste (1867–1936), 79
Clough, G. Wayne, 196

D
Dahler, Hans (born 1876), 19
Dallmann, Eduard (1830–1896), 4, 222, 269
de Geer, Gerard (1858–1943), 83, 85
de Gerlache de Gomery, Adrien (1866–1934), 6
Detmers, Erwin (1889–1912), 142
Diederich, Klaus (gest. 1969), 197
Dreyer, Gustav, 83, 116
Drygalski, Clara (née Wallach, 1883–1958), 9
Drygalski, Erich von (1865–1949), vii, 2, 5–10, 13, 18, 21, 33, 42, 48, 51, 66, 73, 76, 85, 87, 89, 91, 96, 98, 101, 103, 105, 107, 114, 127, 128, 141, 152, 177, 185, 187, 189, 236, 241, 242, 246–251, 254–256, 258, 263, 264, 269, 270

© Springer Nature Switzerland AG 2021
C. Lüdecke, *Germans in the Antarctic*,
https://doi.org/10.1007/978-3-030-40924-1

E

Ellsworth, Lincoln (1880–1951), 163
Elzinga, Aant, 196
Engemann, Fritz, 83
Enzensperger, Josef (1873–1903), 37, 39, 41, 66, 69, 90
Ernst II, Duke of Saxe-Altenburg (1871–1955), 114, 116
Evensen, Carl Julius (1852–1927), 81

F

Federoff, Nina, 196
Filchner, Ilse (née Ostermaier), 76
Filchner, Wilhelm (1877–1957), 76, 79, 81, 83, 86, 91, 92, 102, 109, 129, 243, 250, 252, 263
Fisch, Max (born 1875), 19
Frantzius, Ernst von, 15
Franz, Karl, 3, 6, 19, 73, 152, 165
Frémery, Hermann von, 129
Fuchs, Vivian (1908–1999), 195

G

Gazert, Hans (1870–1961), 10, 25, 42–46, 48, 52–56, 58, 59, 63, 64, 71, 89, 90, 92, 242, 251
Gburek, Leo (1910–1941), 152, 170, 180
Gernandt, Hartwig, 197, 198
Goeldel, Wilhelm von (born 1881), 89, 90, 98, 99, 103, 105, 109, 111, 114, 116, 117, 121, 122, 125, 127, 251
Göring, Hermann (1893–1946), 136, 138, 140, 145, 153, 180, 181, 241, 271
Graser, Nora, 226
Grotewahl, Max (1894–1958), 129–131

H

Hallstein, Walter Peter (1901–1982), 193
Hampel, Herbert, 212
Heim, Fritz (1887–1980), 88, 90, 103, 105, 109, 111, 114, 116, 117, 120, 121, 125–128
Heinacker, Paul (born 1882), 19
Heinrich, Willy (born 1878), 92
Hempel, Gotthilf, 197, 203, 209, 211, 214, 218, 221
Herold, Werner, 210, 211
Herrligkoffer, Karl Maria (1916–1991), 193, 195, 202
Herrmann, Ernst (1895–1970), 134, 146, 152, 155, 163–165, 167, 170, 172, 173, 175, 178, 180–182, 184, 185, 271

Heyneck, Conrad, 117, 125, 127
Hitler, Adolf (1889–1945), 187, 188, 259, 260
Hoel, Adolf (1879–1965), 162, 163
Hoffmann, Kurt, 83

J

Jackson, Frederick (1860–1938), 6
Jacobsen, Marie, 196
Johannsen, Daniel (born 1873), 19, 48, 58
Johansen, Hjalmar (1867–1913), 6
Johnsen, Johann Ludwig, 83
Jupitz, Rudolf, 130

K

Kaufmann, Karl (1900–1969), 2, 182
Kling, Alfred (born 1882), 96, 98, 104, 109, 111, 114, 117, 120, 125, 127
Klück, Karl (born 1869), 19, 58, 83, 92
Knoop, Detlef, 211
Kobarf, Wolfgang, 211
Koch, Lauge (1892–1964), 152
Koch, Robert (1843–1910), 10
Kohl (later Kohl-Larsen), Ludwig (1884–1869), 89, 93, 96, 98, 100–103, 105
Kohnen, Heinz (1938–1997), 195, 201, 202, 207, 208, 220, 225, 227, 229, 230
König, Felix (1880–1945), 76, 90, 92, 98–100, 102, 103, 109, 111, 117, 120, 122, 125, 127
Korhammer, Susanne, 200, 206, 211, 216, 218
Kottas, Alfred (1885–1969), 149, 150
Kraul, Otto (1892–1972), 136, 149, 160, 172
Krause, Franz, 83
Kroogmann, Carl Vincent, 182
Krümmel, Otto (1854–1912), 25, 26

L

Lang, Michael A., 196
Lange, Heinz (1908–1943), 152, 164
Larsen, Carl Anton (1860–1924), 101, 102, 125, 136
Larsen, Margit, 105
Lerche, Wilhelm (born 1864), 27, 50, 58, 63
Lerchenfeld-Koefering, Graf Hugo von und zu (1843–1925), 87
Lewald, Theodor (1860–1947), 79
Lohmann, Hans (1863–1934), 89, 93, 98
Lorenzen, Wilhelm (gest. 1914), 91, 92, 100, 111, 117, 122, 125–127
Lüdecke, Cornelia, 21, 30, 40–42, 72, 88, 105, 190, 196, 213

Luitpold, Bayerischer Prinzregent
(1821–1912), 86, 87, 108, 110,
111, 115
Luther, Hans (1879–1963), 140, 143
Luyken, Karl (1874–1947), 22, 37, 39
Lysell, Wilhelm (born 1873), 19

M

Marek, Reinhold (born 1871), 19
Markham, Clements Sir (1830–1916), 5, 20, 21
Marloth, Rudolf (1855–1931), 37
Matthöfer, Hans Hermann (1925–2009), 202
Mayr, Rudolf (1910–1991), 142, 148, 150,
152, 164, 166–168, 173, 175
Meier, Siegfried (born 1937), 197, 198
Mentzel, Rudolf (1900–1987), 182
Michael, Reinhold (born 1876), 19
Miller, Scott, 196
Möbius, Richard, 212
Moeser, Walter (gest. 1912), 142
Moltke, Graf Helmut von (1848–1916), 79
Muhle, Grazyna, 216, 218
Muhle, Heiko, 211
Müller, Ernst, 83
Müller I, Josef (born 1877), 54, 56
Müller II, Leonard (geb. 1858), 19
Müller, Johannes (1885–1943), 19, 92, 100,
103, 112, 122, 125, 126

N

Nansen, Fridtjof (1861–1930), 6, 16,
25, 85, 86
Neuberger, Kaspar, 88, 92, 98
Neuhaus, C., Kapitän (geb. 1862), 66
Neumayer, Georg von (1826–1909), 3–6, 10,
12, 14, 25, 32, 44, 73, 200, 202–211,
213, 216–219, 222–226, 228–234
Noack, Georg Richard (geb. 1877), 28, 92
Noack, Gerold, 19, 57, 83, 92, 221
Nordenskjöld, Otto (1869–1928), 22, 23, 79,
86, 101, 116

O

Obleitner, Friedrich, 203
Olaisen, Morten, 83, 109
Olsen, Karl Anton, 83
Orheim, Olav, 162, 196
Ott, Ludwig (geb. 1876), 25, 53, 58, 63

P

Paulsen, Karl-Heinz (1909–1941), 152, 177

Penck, Albrecht (1858–1945), 79, 86, 127
Petermann, August (1822–1878), 3, 25
Philipp, Hans (1878–1944), 83, 85, 86
Philippi, Emil (1871–1910), 25, 27, 44, 48, 50,
56, 58, 63, 71
Poland, Willem (died 2008), 229
Popp, Christian (1928–1960), 197
Posadowsky-Wehner, Graf Arthur von
(1845–1932), 12–13, 29, 31, 69, 73, 74
Possin, Albert (geb. 1879), 19
Potpeschnigg, Karl (geb. 1875), 85, 86
Preuschoff, Franz, 152, 165
Prince Heinrich of Prussia (1862–1929),
71, 81, 91
Przybyllok, Erich (1880–1954), 85–87, 90, 96,
97, 100, 102, 105, 114, 115, 117, 120,
122, 127
Puskeppeleit, Monika, 206, 216, 217, 220

R

Rau, Walter (1874–1940), 134, 135
Rave, Christopher (1881–1933), 142
Regula, Herbert (1910–1980), 152, 160,
163, 164
Reimers, August (geb. 1876), 19
Reuterskjöld, Lenhart (geb. 1882), 19, 46, 58
Richardson, Michael, 196
Richter, Max (1856–1921), 40
Richthofen, Ferdinand Freiherr von
(1833–1905), 7, 8, 20, 76
Rintoul, Stephen, 196
Ritscher, Alfred (1879–1963), 129–131,
138–145, 148, 149, 151–156, 158, 160,
162–164, 166–168, 172, 173, 175, 177,
178, 180–187, 191, 193, 199, 213,
214, 221
Ritscher, Ilse (née Uhlmann, 1916–1997), 131,
143, 166, 186
Ritscher, Susanne (née Loewenthal,
1886–1975), 143
Röbke, Karl-Heinz (geb. 1910), 156
Rüdiger, Hermann (1889–1946), vii, 141, 142
Ruser, Hans (1863–1930), 24, 27, 41, 44, 46,
48, 50, 58, 63, 68
Rust, Bernhard (1883–1945), 182

S

Sally, R. Tucker, 196
Sandleben, August, 142
Sauter, Siegfried (1916–2008), 161, 165, 188
Schacht, Hjalmar (1877–1970), 134
Schäfer, Ernst (1910–1992), 156
Schalitz, Louis, 83

Schavan, Annette, 229
Schirmacher, Richardheinrich, 148, 151, 165–167, 170, 175
Schlosser, Elisabeth, 216, 217
Schmidt, Josef, 207, 212
Schmidt-Ott, Friedrich (1860–1956), 14, 79, 87
Schröder-Stranz, Herbert (1888–1912), 140–142, 156
Schug, Joachim, 211
Schulze, Karl Johannes, 83
Schwabe, Adolf, 83, 112, 116
Schwarz, Wilhelm (geb. 1873), 19
Scott, Robert Falcon (1868–1912), 23, 44, 73, 78, 80, 92
Seelheim, Heinrich (1884–1964), 83, 85, 86, 93, 96, 97
Selle, Hermann, 83
Shackleton, Ernest (1874–1922), 79, 85, 86
Simon, Wilhelm, 83
Skeib, Günter (1919–2012), 196–197
Slossarczyk, Walther (1887–1911), 92, 103–105
Sobiesak, Monika, 216
Stade, Hermann (1867–1932), 8
Stehr, Albert (geb. 1874), 24, 56
Stjernblad, Curt (geb. 1882), 19
Szabo, Ladislas, 187

T

Tafel, Albert (1876–1935), 66, 67, 76, 77, 89, 98
Tirpitz, Alfred von (1849–1930), 1–2, 10, 12, 15
Triggs, Guillan, 196

U

Ule, Willi (1861–1940), 80, 93, 96

V

Vahsel, Richard (1868–1912), 27, 48, 56, 70, 91, 92, 94, 96, 98, 100, 101, 108–111, 114–117, 121–123
Vanhöffen, Ernst (1858–1918), 8, 24, 27, 29, 35, 37, 46, 53, 54, 56, 74
Victoria Empress and Queen of Prussia (1840–1901), 32

W

Wallner, Klaus (died 1983), 210
Walton, R. David, 196
Weddell, James, 108
Wegener, Alfred (1880–1930), 122, 202, 225
Weigel, Ursula, 216
Weigelt, Estella, 216
Werth, Emil (1869–1958), 27, 37, 41
Wienke, Georg, 28, 37
Wilhelm II, Kaiser (1859–1941), 27, 28, 73, 79
Wilken, Johann, 83, 100
Winter, Stefan (died 2008), 229
Woerner, Elisabeth, 129
Wohlthat, Helmuth (1893–1982), 134, 136, 138, 139, 141, 145, 151, 153, 155, 160, 163, 173, 180, 182, 184–186, 190, 200
Wolff, Paul, 83
Wyputta, Ulrike, 216

Z

Zäncker, Karl, 83

MIX
Papier aus verantwortungsvollen Quellen
Paper from responsible sources
FSC® C105338

If you have any concerns about our products,
you can contact us on
ProductSafety@springernature.com

In case Publisher is established outside the EU,
the EU authorized representative is:
**Springer Nature Customer Service Center GmbH
Europaplatz 3, 69115 Heidelberg, Germany**

Printed by Libri Plureos GmbH
in Hamburg, Germany